Chapter **3** Continued

Chapter **4**

THE MINOR ELECTRICAL INSTALLATION WORKS CERTIFICATE

Chapter **5**

THE PERIODIC INSPECTION REPORT

Contents

Chapter **5** Continued

Chapter **6** Continued

THE SCHEDULES

Inspection, Testing and Certification
including Periodic Reporting

PRACTICAL ADVICE AND GUIDANCE

The NICEIC

The National Inspection Council for Electrical Installation Contracting is a non-profit making organisation set up to protect electricity users against the hazards of unsafe and unsound electrical installations. The NICEIC is supported by all sectors of the electrical industry, consumer bodies, and local and central Government.

The NICEIC is the industry's independent voluntary regulatory body on electrical safety matters.

Published by

NICEIC
Vintage House
37 Albert Embankment
London SE1 7UJ

ISBN 0-9531058-3-0

Contents

Chapter 1

INTRODUCTION

Chapter 2

THE ELECTRICAL INSTALLATION COMPLETION CERTIFICATE

Chapter **2** Continued

Chapter **3**

THE DOMESTIC ELECTRICAL INSTALLATION CERTIFICATE

Chapter 7

INSPECTION

Chapter Continued

Chapter **8**

TESTING

Contents

Chapter 8 Continued

This second edition of the NICEIC book on inspection, testing, certification and reporting has been fully revised. It takes into account, amongst other things, the changes to regulatory requirements arising from the second amendment to *BS 7671: 1992*, published in December 1997, including the revision of the model forms of certification and reporting.

January 2000

1

Introduction

Chapter 1

 CHAPTER 1 INTRODUCTION

AIMS AND OBJECTIVES

Even the most experienced electrical engineers and electricians can be expected to make the occasional mistake when carrying out electrical installation work. This is one of the main reasons why every new electrical installation, and every alteration and addition to an existing installation, must be thoroughly inspected and tested, and any defects or omissions found made good, before the installation, or the alteration or addition, is put into service. The other key reason for inspection and testing is to provide a record of the condition of a new installation when it is first put into service, or when an altered or extended installation is restored to service.

All electrical equipment deteriorates with age, as well as with wear and tear from use. Every electrical installation therefore needs to be inspected and tested at appropriate intervals during its lifetime to establish that its condition is such that, subject to the completion of any necessary remedial work, the installation is safe to remain in service at least until the next inspection is due.

Correctly compiled Electrical Installation Certificates, Domestic Electrical Installation Certificates, Minor Electrical Installation Works Certificates and Periodic Inspection Reports provide the persons responsible for the safety of electrical installations (including contractors, owners and users) with an important record of the condition of those installations at the time they were inspected and tested. Such certificates and reports also provide an essential basis for subsequent inspection and testing, without which a degree of costly exploratory work might be necessary on each occasion. In the event of injury or fire alleged to have been caused by an electrical installation, certificates and reports will provide documentary evidence to help demonstrate that, in the opinion of competent persons, the installation had been installed and subsequently maintained to a satisfactory standard of safety.

Every electrical contractor should employ at least one experienced person who has responsibility for inspecting and testing electrical installation work in accordance with the requirements of the national standard for electrical safety, *BS 7671:1992 Requirements for Electrical Installations* (otherwise known as the *IEE Wiring Regulations*). This includes the preparation, to a satisfactory standard, of the forms of certification and reporting associated with inspection and testing, in order to accurately record the results for the benefit of both the contractor and the users.

The aim of this book is to promote good practice by providing Approved Contractors and others with practical advice and guidance which answers many of the questions commonly arising during the inspection and testing of electrical installation work, or during the preparation of the associated certificates and reports.

NICEIC certificates and reports used for inspection and testing

Versions of the four key certificates and forms, coloured green, are available for use by electrical contractors not enrolled with NICEIC, in particular those preparing for enrolment.

Extracts from *Rules and Regulations for the Prevention of Fire Risks Arising from Electric Lighting* published by the Society of Telegraph Engineers and of Electricians, 11 May, 1882.

"The difficulties that beset the electrical engineer are chiefly internal and invisible, and they can only be guarded against by 'testing,' or probing with electric currents." (Introduction)

"NB - The value of frequently testing the wires cannot be too strongly urged. It is an operation, skill in which is easily acquired and applied. The escape of electricity cannot be detected by the sense of smell, as can gas, but it can be detected by apparatus far more certain and delicate. Leakage not only means waste, but in the presence of moisture it means destruction of the conductor and its insulating covering, by electric action." (Note to Rule 17)

This second edition of the book updates the guidance given and takes into account changes to the regulatory requirements and the revision of the model forms of certification and reporting arising from the publication, in December 1997, of the second amendment (AMD 9781) to *BS 7671:1992*.

SCOPE

This book is intended to complement Part 7 of *BS 7671* and the information and advice provided in other authoritative publications such as *IEE Guidance Note 3: Inspection and Testing*.

The book covers the general requirements relating to the inspection and testing of electrical installations forming part of TN-C-S, TN-S and TT systems. It does not cover the particular requirements relating to TN-C and IT systems, which are not common in the UK.

The book does not cover the particular requirements relating to the inspection, testing and certification of specialised electrical installations such as fire alarm and emergency lighting systems, or installations in hazardous areas. Nor does it cover in detail the particular requirements relating to non-conducting locations, earth-free equipotential zones, site-applied insulation and the like.

The book is not intended to instruct untrained and inexperienced persons to undertake the inspection and testing of electrical installations. It assumes that all persons proposing to undertake such work have already acquired the necessary knowledge, understanding and skill, and are properly equipped to undertake such work without putting themselves or others at risk.

Persons who consider that they are not fully competent to undertake the inspection and testing of electrical installations without direct supervision should, as a first step, complete an appropriate course and assessment provided by a reputable training organisation.

ARRANGEMENT OF THE BOOK

This book uses as its main theme the preparation of the NICEIC versions of the industry's principal forms of certification and reporting. All the relevant details and declarations relating to the safety of an electrical installation are intended to be recorded on these forms, which consist of:

- The **Electrical Installation Certificate** - intended for new installation work, including alterations and additions to existing installations.

- The **Domestic Electrical Installation Certificate** - intended for new installation work, including alterations and additions to existing installations, in a single dwelling (house or individual flat).

Inspection, Testing, Certification and Reporting

BS 7671: 1992 - Requirements for Electrical Installations, incorporating Amendments 1 and 2.

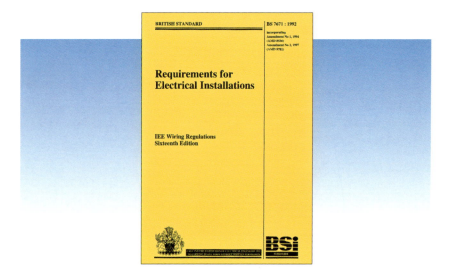

The purpose of inspecting and testing electrical installation work is to check whether it:

- Meets the requirements of *BS 7671* and any other applicable standards.

- Is properly constructed.

- Only incorporates equipment which has been correctly selected for its purpose.

It is in the best interests of both the contractor and the user that inspection, testing, certification and reporting is carried out correctly.

- The **Minor Electrical Installation Works Certificate** - intended for minor installation work that does not include the provision of a new circuit.

- The **Periodic Inspection Report for an Electrical Installation** - intended for reporting on the condition of an existing installation.

The particular versions of the forms on which this book is based are those published in January 2000 for use by NICEIC Approved Contractors.

However, the guidance is equally applicable to other forms of certification and reporting based on the model forms given in BS 7671, especially the versions published by the NICEIC in January 2000 for use by non-approved contractors. (See page 4).

Some of the inspection and testing procedures required for periodic inspection reporting are common to those required to verify new installation work. To avoid duplication, such procedures are covered in the book only once, the reader being referred to another chapter where appropriate.

THE PURPOSE OF INSPECTION, TESTING AND CERTIFICATION OR REPORTING

The Wiring Regulations date back to 1882. The first edition, called the *Rules and Regulations for the Prevention of Fire Risks Arising from Electric Lighting*, was the first in a long series published by the Institution of Electrical Engineers (IEE) and its predecessor, the Society of Telegraph Engineers and of Electricians. The importance of testing was recognised even in the earliest days, as can be seen from the extracts shown on page 4.

The sentiments expressed in those early rules and regulations are as significant today as they were in 1882.

The fundamental reason for inspecting and testing an electrical installation is to determine whether new installation work is safe to be put into service, or whether an existing installation is safe to remain in service until the next inspection is due.

If it is verified that new installation work fully complies with the current requirements of *BS 7671*, it will have been correctly designed and constructed, all equipment will be suitable for its intended purpose and, consequently, the installation will be safe to be put into service. Verification includes inspection and testing during construction and on completion, and comparison of the results with the requirements of *BS 7671* and with any particular electrical safety requirements detailed in the design information. The process of verification is intended to ensure that the electrical installation is safe to put into service, and that it is likely to remain in such a condition at least until the first periodic inspection is due.

HSE Memorandum of Guidance on the Electricity at Work Regulations 1989

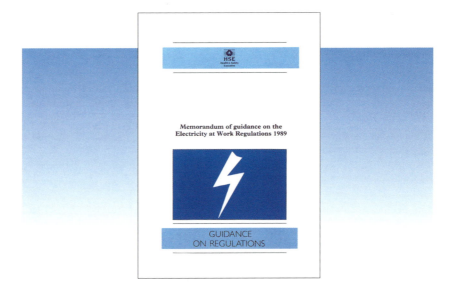

Correctly compiled certificates and reports:

- Are a record of your involvement and responsibility.

- Demonstrate that you have carried out the necessary inspection and testing.

- Can be significant in providing you, if necessary, with a defence under Regulation 29 of the *Electricity at Work Regulations.*

NOTE: In Northern Ireland, the *Electricity at Work Regulations (Northern Ireland) 1991* apply, and there is a corresponding Memorandum of Guidance issued by the Health and Safety Agency for Northern Ireland. The Regulations and the Memorandum of Guidance are similar to those applicable in other parts of the UK.

Confirmation that a new installation is safe for use is provided by the issue of a correctly compiled Electrical Installation Certificate or Domestic Electrical Installation Certificate to the person who ordered the work. For an alteration or addition not extending to the provision of a new circuit, a Minor Electrical Installation Works Certificate may be used instead of an Electrical Installation Certificate or a Domestic Electrical Installation Certificate. Any defects or omissions found during the process of verification must be corrected (and inspection and tests repeated as necessary) before the certificate is issued.

Having been certificated and put into service, all electrical installations need to be inspected and tested at appropriate intervals throughout their lifetime. On each occasion, details of the condition of the installation should be recorded in a Periodic Inspection Report for the benefit of the person ordering the inspection, and of the persons subsequently involved in remedial work or further inspections.

Certificates and reports provide an important and valuable record of the condition of an electrical installation at the time of the inspecting and testing, together with details of any extension or alteration work which may have been carried out since the installation was first put into service.

LEGAL ASPECTS

Whilst the *Electricity at Work Regulations* do not specifically require electrical installation certificates or reports to be issued and retained, such documents may provide the only effective evidence for the contractor or the person(s) responsible for the safety of an electrical installation should either or both be prosecuted under the provisions of those statutory regulations.

All persons carrying out inspection and testing of electrical installations, whether as an employee or self-employed person, must comply with the relevant requirements of the *Electricity at Work Regulations*. They apply to all persons carrying out any work activity falling within the scope of the Regulations. They therefore apply to inspection and testing (as well as to other aspects of electrical work) even when being carried out in locations not normally considered to be places of work, such as domestic premises.

Persons undertaking inspection and testing should pay particular attention to Regulation 14, 'working on or near live conductors', and to the associated guidance provided in the *Memorandum of Guidance on the Electricity at Work Regulations*.

Electrical contractors that issue inaccurate or misleading certificates or reports are liable to criminal prosecution by Trading Standards Officers under the Trade Descriptions Act 1968 or other relevant legislation. There have also been successful claims for damages in the civil courts by persons who have relied upon the content of an electrical installation certificate or report, such as for house purchase.

The importance of competence and safe working procedure

Everyone may occasionally be tempted to cut corners to save time or money, but the personal, legal and financial consequences of taking such risks can be severe, especially where inspection and testing is concerned. The following is based on an article that appeared in the technical press in 1997:

An enquiry into an electrical accident in which a man was temporarily blinded revealed that unqualified workers were being used by an electrical contractor to carry out electrical testing work.

The injured electrician, who was carrying out a continuity test on a sub-main cable, failed to check that two phase conductor terminals were dead before he began work. In fact the terminals were live, with 400 V between them. A flash of electric current which engulfed his body causing electrical burns to his hands and face, and temporarily blinding him, occurred when he attempted to put a length of copper wire across the terminals.

On interviewing the injured electrician, Local Authority Inspectors found that he had no formal qualifications relating to electrical work, and that all his knowledge was gained from working with other electricians. He had little experience of testing work, and did not fully understand the tests being carried out.

The electrical contractor was charged with three breaches of Section 2 of the Health and Safety at Work etc Act 1974, for each of which they were fined £12,000. The company was also fined £3,000 for a breach of the Electricity at Work Regulations 1989.

Based on an article published in the August 1997 issue of The Safety and Health Practitioner.

Such prosecutions can be avoided by employing only competent persons to produce the certificates and reports, and by agreeing the exact scope of work with the client before commencing.

THE INSPECTOR

Throughout this book, the term 'inspector' is used to describe a person responsible for inspecting and testing an electrical installation. All persons carrying out the inspection and testing of electrical installations must be competent to do so, unless under the direct supervision of a competent person.

The NICEIC considers that, to be competent to undertake the inspection and testing of an electrical installation, persons must as a minimum:

- Have sufficient knowledge and experience of electrical installation matters to avoid danger to themselves and to others.

- Be familiar with, and understand, the requirements of *BS 7671*, including those relating to inspection, testing, certification and reporting.

- Have a sound knowledge of the particular type of installation to be inspected and tested.

- Have sufficient information about the function and construction of the installation to allow them to proceed in safety.

If the inspector is competent and takes all the necessary safety precautions, including following the correct procedure, the process of inspection and testing should not create danger to persons or livestock, or cause damage to property.

NICEIC experience indicates that persons undertaking periodic inspection reporting need to have above-average knowledge and experience of electrical installation matters to enable them to safely and accurately assess the condition of an existing electrical installation, especially when they do not have access to the design information relating to that installation.

Safe working procedure

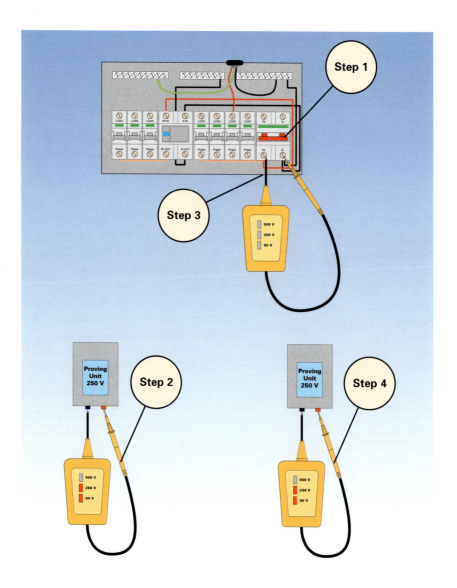

SAFE WORKING PROCEDURE

Isolation

Regulation 712-01-01 of *BS 7671: 1992* states that 'inspection shall precede testing and shall normally be done with that part of the installation under inspection disconnected from the supply'. The appropriate form of disconnection is isolation.

Isolation is not simply the operation of an isolating device. It is a procedure, the objective of which is to ensure that the supply is cut off and remains cut off for reasons of safety. The process of isolation can be broken down into a number of steps as necessary for the particular circumstances. As a minimum, the following safe working procedure should be adopted before commencing inspection:

Step 1

Open the means of isolation for the circuit(s) to be inspected (having first removed the load, if the isolator is an off-load device), and secure the isolating device in the open position with a lock or other suitable means.

Step 2

Prove the correct operation of a suitable voltage test instrument against a known source.

Step 3

Test, with the voltage test instrument, the circuit(s) to be worked on to verify that no dangerous voltage is present.

Step 4

Prove the voltage test instrument again against the known source to check that it was functioning correctly when the circuit(s) were tested for the presence of voltage.

HSE Guidance Note GS 38 - Electrical Test Equipment for use by Electricians

HSE Guidance Note GS 38

The HSE publication gives guidance on the safe use of electrical test equipment.

Measurement of voltage and current

Particular care should be taken when measuring voltage, prospective fault current and earth fault loop impedance, as the tests have to be conducted when the installation is energised. Take appropriate precautions to prevent accidents, such as ensuring that test instruments and test leads are suitable and properly used, that the equipment to be worked upon is safe for the intended tests, and that the working environment does not present additional dangers *(see HSE Guidance Note GS 38)*.

TEST EQUIPMENT

HSE Guidance Note GS 38: Electrical Test Equipment for use by Electricians gives advice to competent persons on the selection and use of test equipment for circuits with rated voltages not exceeding 650 V. Those involved in electrical inspection and testing should familiarise themselves with the guidance given in this publication.

Aspects that should be checked with regard to test equipment include the following (refer to the HSE Guidance Note for full information):

- Test instruments and their associated test leads must be suitably constructed and fit for the purpose for which they are to be used. Such equipment should preferably meet the safety requirements set out in *BS EN 61010* and the performance standards detailed in the appropriate Part(s) of *BS EN 61557*.

- Test equipment should be checked for suitable rating for the anticipated voltage prior to testing any circuit which is, or may be, energised.

- Test instruments and their associated test leads should be checked before each use for signs of damage and deterioration, and be maintained such that they remain safe for use in the intended manner.

- Test instruments and test leads should be used only in the environment(s) for which they have been designed.

- Test leads should have shrouded terminals and finger guards, and fuses or current-limiting resistors. Where an instrument and its test leads comply with product specification *BS EN 61010* and with the appropriate Parts of *BS EN 61557*, and where the instrument manufacturer's written user instructions make clear that the instrument may be so used, the requirement for test leads to incorporate fuses or current-limiting resistors may be relaxed.

- Probes should have no more than 4 mm of exposed metal tip. (It is strongly recommended that the exposed metal tip does not exceed 2 mm).

Test probes and leads

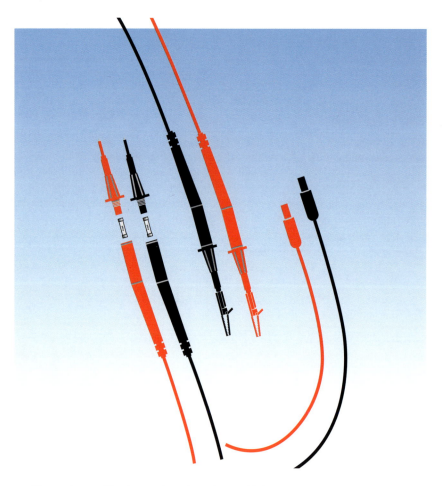

Test probes and leads must be selected to avoid danger.

ASSESSMENT OF INSPECTION AND TEST RESULTS

It is not sufficient for the inspector simply to record the results of inspection and testing on a certificate or report. The inspector should first consider and assess the results to determine whether they are as expected for the particular type of installation, and that they are consistent with installation work complying with all the relevant requirements of *BS 7671*.

For example, it would not be acceptable for the inspector to issue an Electrical Installation Certificate or a Domestic Electrical Installation Certificate recording a measured earth fault loop impedance value for a circuit in a new installation which exceeded the limiting value permitted by Regulation 413-02-08 (taking account of temperature rise under fault conditions) for disconnection of the protective device within the appropriate time limit to provide protection against indirect contact. Such an excessive impedance would need to be investigated, appropriate remedial action taken, and the test repeated before the Electrical Installation Certificate or Domestic Electrical Installation Certificate was issued and the installation put into service.

If an excessive earth fault loop impedance is discovered during the course of a periodic inspection of an existing installation, the measured (excessive) value should be recorded, together with an appropriate observation and recommendation for remedial action, on the Periodic Inspection Report.

COMPUTER-ASSISTED PREPARATION OF CERTIFICATES AND REPORTS

Whilst the use of personal computers to assist with the compilation of certificates and reports, perhaps in conjunction with data-logging test instruments, is not a substitute for a competent person having the necessary knowledge, skill and experience, such use may improve the clarity and presentation of the completed certificates and reports.

Some software packages also provide an element of validation or checking of the inspection and test results, identifying obvious errors, omissions and other problems in the data for consideration by a competent person, before the details are printed on to the certificate or report. Such software packages, correctly used, can be expected to help ensure the completeness and accuracy of certificates and reports.

The principal certificate and report forms published by the NICEIC are designed to be completed by hand or computer printer.

Where the information provided on an NICEIC certificate or a report has been checked, compiled and printed with the aid of a proprietary software package endorsed by the NICEIC, a computer-printed copy of the NICEIC logo will appear in the box at the top of the front page of the certificate or report and on the schedule(s) of test results, together with an indication of the supplier of the software and the software version number.

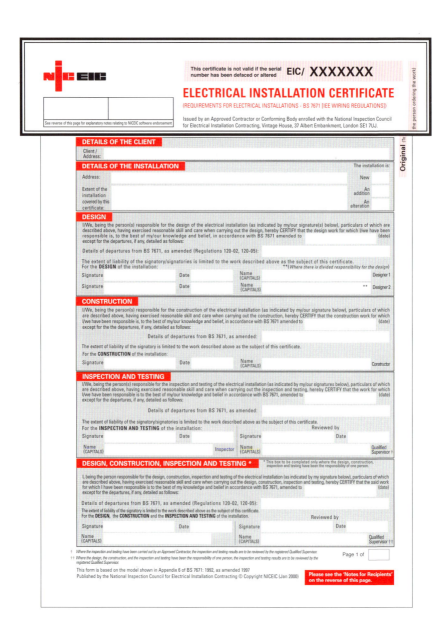

If no computer-printed copy of the NICEIC logo appears in the boxes, the technical information provided on the certificate or report has not been subjected to the system of automatic checks endorsed by the NICEIC.

Where available for the Electrical Installation Certificates, Domestic Electrical Installation Certificates and Minor Electrical Installation Works Certificates published by the NICEIC, NICEIC-endorsed software embodies a system of checking the completeness and acceptability of the inspection and test results.

Irrespective of whether or not a certificate is completed with the aid of software, it remains the responsibility of the compiler of the certificate to ensure that the information provided on the certificate is factual and confirms that the electrical installation work to which the certificate relates is safe to be put into service.

As a Periodic Inspection Report is intended to provide an objective assessment of the condition of an existing installation, NICEIC-endorsed software for such reports, where available, is designed to draw the attention of the compiler to any test results which do not meet the requirements of the current issue of *BS 7671*. However, unlike the software for certifying new work, the software will then permit those test results to be printed on to the report form to provide an accurate record. The compiler of the report is nevertheless expected to include an appropriate observation and recommendation for remedial action for each such item of non-compliance.

Irrespective of whether or not a periodic inspection report is completed with the aid of software, it remains the responsibility of the compiler of the report to ensure that the information provided in the report is factual and accurately records the condition of the electrical installation to which the report relates, having regard to the stated extent and limitations of the report.

SERIAL NUMBERS ON NICEIC CERTIFICATES AND REPORTS

A unique serial number (including prefix) is printed on the original and duplicate of each NICEIC certificate and report form published for use by Approved Contractors. The serial number allows the origin of the certificate or report to be traced by the NICEIC in the event of a query as to its authenticity. For this reason, no organisation other than the Approved Contractor to which an NICEIC certificate or report form has been supplied by the NICEIC is authorised to issue that form. If the serial number has been erased or altered in any way, the NICEIC would consider the certificate or report to be invalid.

The NICEIC strongly prefers Approved Contractors to issue NICEIC certificates and report forms, as this provides a measure of confidence to the recipients. Non-approved contractors are not authorised to issue NICEIC certificates or reports, but similar forms conforming to the national standard are available from the NICEIC for use by electrical contractors not enrolled with the NICEIC. These versions have no serial numbers.

Inspection, Testing, Certification and Reporting

2

The Electrical Installation Certificate

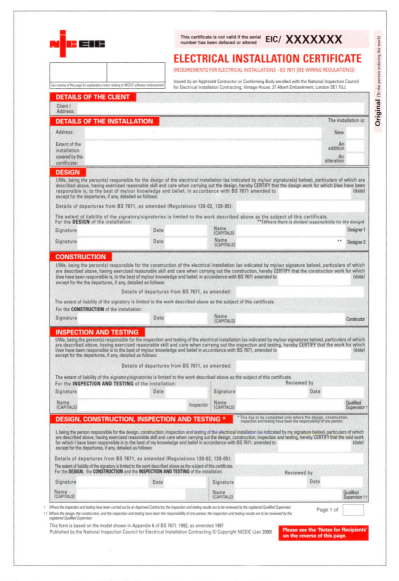

The front page of the NICIEC Electrical Installation Certificate.

THE ELECTRICAL INSTALLATION CERTIFICATE

REQUIREMENTS

BS 7671 requires a standard form of Electrical Installation Certificate (including a schedule of test results) to be issued for all new installation work, and for all alterations and additions to existing installations. A separate certificate must be issued for each distinct electrical installation. The certificate provides a formal assurance from those responsible for the design, construction, and inspection and testing of the installation that it complies with the requirements of the national standard for electrical safety.

A certificate (based on the model form given in *BS 7671*) must be issued to the person who ordered the work, whether or not the contractor is registered with the NICEIC, and whether or not a certificate has been specifically requested by the client. The NICEIC strongly prefers Approved Contractors to issue NICEIC certificates, as this provides a measure of confidence to the recipients. Non-approved contractors are not authorised to issue NICEIC certificates, but similar certificates based on the model form given in *BS 7671* are available from the NICEIC.

If the contractor responsible for the construction of the installation is not also responsible for its design and/or inspection and testing, it will be necessary for the contractor to liaise with the other parties responsible for those aspects in order to complete all the details required on the certificate.

It should be noted that the Electrical Installation Certificate is not intended to be issued to confirm the completion of a contract. The certificate is a declaration of electrical safety which should be issued before an installation is put into service. The certificate should not be withheld for contractual reasons, especially if the electrical installation is available for use.

The Electrical Installation Certificate is to be used only for the initial certification of a new installation, or of new work associated with an alteration or addition to an existing installation, carried out in accordance with *BS 7671: 1992* as amended. Alternatively, where the electrical installation work relates to a single dwelling (house or individual flat) a Domestic Electrical Installation Certificate may be used, provided the installation falls within the scope of that certificate (see Chapter 3). Neither certificate is to be used for a periodic inspection of an existing installation, for which a Periodic Inspection Report form should be used (see Chapter 5).

This certificate is not valid if the serial number has been defaced or altered

EIC/ **XXXXXXX**

Original (To the person ordering the work)

NIC EIC

See reverse of this page for explanatory notes relating to NICEIC software endorsement

ELECTRICAL INSTALLATION CERTIFICATE
(REQUIREMENTS FOR ELECTRICAL INSTALLATIONS - BS 7671 [IEE WIRING REGULATIONS])

Issued by an Approved Contractor or Conforming Body enrolled with the National Inspection Council for Electrical Installation Contracting, Vintage House, 37 Albert Embankment, London SE1 7UJ.

DETAILS OF THE CLIENT

Client / Address:

DETAILS OF THE INSTALLATION

The installation is:

Address:

Extent of the installation covered by this certificate:

New

An addition

An alteration

DESIGN

I/We, being the person(s) responsible for the design of the electrical installation (as indicated by my/our signature(s) below), particulars of which are described above, having exercised reasonable skill and care when carrying out the design, hereby CERTIFY that the design work for which I/we have been responsible is, to the best of my/our knowledge and belief, in accordance with BS 7671 amended to ___ (date) except for the departures, if any, detailed as follows:

Details of departures from BS 7671, as amended (Regulations 120-02, 120-05):

The extent of liability of the signatory/signatories is limited to the work described above as the subject of this certificate.
For the **DESIGN** of the installation: ******(Where there is divided responsibility for the design)

| Signature | Date | Name (CAPITALS) | | Designer 1 |
| Signature | Date | Name (CAPITALS) | ** | Designer 2 |

CONSTRUCTION

I/We, being the person(s) responsible for the construction of the electrical installation (as indicated by my/our signature below), particulars of which are described above, having exercised reasonable skill and care when carrying out the construction, hereby CERTIFY that the construction work for which I/we have been responsible is, to the best of my/our knowledge and belief, in accordance with BS 7671 amended to ___ (date) except for the departures, if any, detailed as follows:

Details of departures from BS 7671, as amended:

The extent of liability of the signatory is limited to the work described above as the subject of this certificate.
For the **CONSTRUCTION** of the installation:

| Signature | Date | Name (CAPITALS) | Constructor |

INSPECTION AND TESTING

I/We, being the person(s) responsible for the inspection and testing of the electrical installation (as indicated by my/our signatures below), particulars of which are described above, having exercised reasonable skill and care when carrying out the inspection and testing, hereby CERTIFY that the work for which I/we have been responsible is to the best of my/our knowledge and belief in accordance with BS 7671, amended to ___ (date) except for the departures, if any, detailed as follows:

Details of departures from BS 7671, as amended:

The extent of liability of the signatory/signatories is limited to the work described above as the subject of this certificate.
For the **INSPECTION AND TESTING** of the installation: Reviewed by

| Signature | Date | Signature | Date | |
| Name (CAPITALS) | Inspector | Name (CAPITALS) | | Qualified Supervisor † |

DESIGN, CONSTRUCTION, INSPECTION AND TESTING *

* This box to be completed only where the design, construction, inspection and testing have been the responsibility of one person.

I, being the person responsible for the design, construction, inspection and testing of the electrical installation (as indicated by my signature below), particulars of which are described above, having exercised reasonable skill and care when carrying out the design, construction, inspection and testing, hereby CERTIFY that the said work for which I have been responsible is to the best of my knowledge and belief in accordance with BS 7671, amended to ___ (date) except for the departures, if any, detailed as follows:

Details of departures from BS 7671, as amended (Regulations 120-02, 120-05):

The extent of liability of the signatory is limited to the work described above as the subject of this certificate.
For the **DESIGN**, the **CONSTRUCTION** and the **INSPECTION AND TESTING** of the installation. Reviewed by

| Signature | Date | Signature | Date | |
| Name (CAPITALS) | | Name (CAPITALS) | | Qualified Supervisor †† |

† Where the inspection and testing have been carried out by an Approved Contractor, the inspection and testing results are to be reviewed by the registered Qualified Supervisor.

†† Where the design, the construction, and the inspection and testing have been the responsibility of one person, the inspection and testing results are to be reviewed by the registered Qualified Supervisor.

This form is based on the model shown in Appendix 6 of BS 7671: 1992, as amended 1997
Published by the National Inspection Council for Electrical Installation Contracting © Copyright NICEIC (Jan 2000)

Page 1 of

Please see the 'Notes for Recipients' on the reverse of this page.

Each NICEIC certificate has a unique serial number to provide traceability.

Where a certificate is to be issued for an alteration or addition to an existing installation, the designer is required to ascertain that the rating and condition of any existing equipment, including that of the supplier (which may have to carry any additional load), are adequate to accommodate in safety the altered circumstances resulting from the modifications, and that the earthing and bonding arrangements are also adequate (see Regulation 130-09-01).

Where an Approved Contractor discovers the existence of a dangerous or potentially dangerous situation in the existing installation (such as the absence of earthing or main bonding where the method of protection against indirect contact is EEBAD), the new work should not proceed and the client should be advised immediately, preferably in writing, to satisfy the duties imposed on competent persons by the *Electricity at Work Regulations 1989*.

An NICEIC Electrical Installation Certificate may be issued only by the Approved Contractor responsible for the installation work. A certificate must never be issued to cover another contractor's installation work.

An NICEIC Electrical Installation Certificate which has pages missing is considered invalid. The NICEIC certificate has at least five pages, depending on the number and type of circuits in the installation. For installations having more than one distribution board or more circuits than can be recorded on pages 4 and 5, one or more additional pages of the Schedule of Circuit Details for the Installation and Schedule of Test Results for the Installation will be required.

Continuation schedules are available separately from the NICEIC. The additional pages should be given the same unique serial number as the other pages of the certificate, by first striking out 'PIR' and then adding the remainder of the unique serial number in the space allocated. Superseded versions of the continuation schedules must not be used in conjunction with these certificates. The page number for each additional schedule should be inserted, together with the total number of pages comprising the certificate (eg page 6 of 7).

Generally, irrespective of the method of compilation of the form, all unshaded data-entry boxes should be completed by inserting the necessary text, a 'Yes' or a '✔' to indicate that the task has been completed, or a numeric value of the measured parameter, or by entering 'N/A' meaning 'Not Applicable', where appropriate.

Regardless of the method of compilation of the certificate (software-assisted or by hand), it remains the responsibility of the compiler(s) of the certificate to ensure that the information provided on the certificate is factual, and that the electrical installation work to which the certificate relates is safe to be put into service.

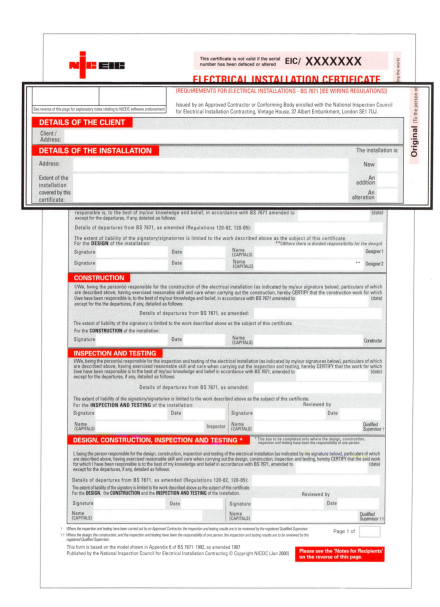

Where the electrical work to which the certificate relates includes the installation of a fire alarm system and/or an emergency lighting system (or a part of such systems) in accordance with British Standards *BS 5839* and *BS 5266* respectively, the Electrical Installation Certificate must be accompanied by a separate certificate or certificates as prescribed by those Standards.

COMPILATION OF THE CERTIFICATE

The remainder of this chapter provides guidance on completing each part of the certificate.

DETAILS OF THE CLIENT

The details to be provided are the name and address of the person or organisation that engaged the contractor to carry out the installation work. Where, for example, the electrical contractor is engaged by a builder, the details entered should be those of the builder. In the case of lengthy name and address details, suitable abbreviations may be used if acceptable to the client.

DETAILS OF THE INSTALLATION

Address

The address to be provided is the complete postal address of the installation, including the postcode.

Extent of the installation covered by this certificate

For a new installation, the certificate may be intended to apply to the whole of the electrical installation at the address given. If not, it is essential to clearly define the extent of the installation to which the certificate applies in terms of areas or services; for example, 'whole electrical installation second floor' or 'whole electrical installation except fire detection and alarm system'.

The description will depend on the particular circumstances, and must be tailored to identify explicitly the extent of the installation work covered by the certificate. In the case of additions or alterations, particular care must be taken to define exactly what work the certificate covers.

The box should be used to identify and clarify any particular exclusions agreed with the client. Any other exclusions which may not be readily apparent to the client or the installation user should also be recorded here.

NICEIC

This certificate is not valid if the serial number has been defaced or altered **EIC/ XXXXXXX**

ELECTRICAL INSTALLATION CERTIFICATE
(REQUIREMENTS FOR ELECTRICAL INSTALLATIONS - BS 7671 [IEE WIRING REGULATIONS])

Issued by an Approved Contractor or Conforming Body enrolled with the National Inspection Council for Electrical Installation Contracting, Vintage House, 37 Albert Embankment, London SE1 7UJ.

See reverse of this page for explanatory notes relating to NICEIC software endorsement

(To the person ordering the work)

DETAILS OF THE CLIENT
Client / Address:

DETAILS OF THE INSTALLATION

The installation is:

New

An addition

An alteration

Original

DESIGN
I/We, being the person(s) responsible for the design of the electrical installation (as indicated by my/our signature(s) described above, having exercised reasonable skill and care when carrying out the design, hereby CERTIFY that the design for which I am/we are responsible is, to the best of my/our knowledge and belief, in accordance with BS 7671 amended to except for the departures, if any, detailed as follows:

Details of departures from BS 7671, as amended (Regulations 120-02, 120-05):

The extent of liability of the signatory/signatories is limited to the work described above as the subject of this certificate.
For the **DESIGN** of the installation: **(Where there is divided responsibility for the design)

| Signature | Date | Name (CAPITALS) | | Designer 1 |
| Signature | Date | Name (CAPITALS) | ** | Designer 2 |

CONSTRUCTION
I/We, being the person(s) responsible for the construction of the electrical installation (as indicated by my/our signature below), particulars of which are described above, having exercised reasonable skill and care when carrying out the construction, hereby CERTIFY that the construction work for which I/we have been responsible is to the best of my/our knowledge and belief, in accordance with BS 7671 amended to   (date) except for the the departures, if any, detailed as follows:

Details of departures from BS 7671, as amended:

The extent of liability of the signatory is limited to the work described above as the subject of this certificate.
For the **CONSTRUCTION** of the installation:

| Signature | Date | Name (CAPITALS) | Constructor |

INSPECTION AND TESTING
I/We, being the person(s) responsible for the inspection and testing of the electrical installation (as indicated by my/our signatures below), particulars of which are described above, having exercised reasonable skill and care when carrying out the inspection and testing, hereby CERTIFY that the work for which I/we have been responsible is to the best of my/our knowledge and belief in accordance with BS 7671, amended to   (date) except for the departures, if any, detailed as follows:

Details of departures from BS 7671, as amended:

The extent of liability of the signatory/signatories is limited to the work described above as the subject of this certificate.
For the **INSPECTION AND TESTING** of the installation: Reviewed by

| Signature | Date | Signature | Date | |
| Name (CAPITALS) | Inspector | Name (CAPITALS) | | Qualified Supervisor † |

DESIGN, CONSTRUCTION, INSPECTION AND TESTING *
* This box to be completed only where the design, construction, inspection and testing have been the responsibility of one person.

I, being the person responsible for the design, construction, inspection and testing of the electrical installation (as indicated by my signature below), particulars of which are described above, having exercised reasonable skill and care when carrying out the design, construction, inspection and testing, hereby CERTIFY that the said work for which I have been responsible is to the best of my knowledge and belief in accordance with BS 7671, amended to   (date) except for the departures, if any, detailed as follows:

Details of departures from BS 7671, as amended (Regulations 120-02, 120-05):

The extent of liability of the signatory is limited to the work described above as the subject of this certificate.
For the **DESIGN**, the **CONSTRUCTION** and the **INSPECTION AND TESTING** of the installation.   Reviewed by

| Signature | Date | Signature | Date | |
| Name (CAPITALS) | | Name (CAPITALS) | | Qualified Supervisor †† |

Please see the 'Notes for Recipients' on the reverse of this page.

If there is insufficient space in the box to give a clear description of the work done and/or any exclusions, then an additional record should be included in the certificate by attaching a separate sheet. Reference to this additional page should be included in the box.

It should be appreciated that failure to clearly describe the extent of the work covered by the certificate could involve the contractor in unforeseen liabilities at a later date.

Nature of installation work

Three boxes are provided to enable the contractor to identify the nature of the installation work:

- 'New' - This box should be ticked only if the whole installation has been installed as new, or if a complete rewire has been carried out.

- 'An addition' (to an existing installation) - This box should be ticked if an existing installation has been modified by the addition of one or more new circuits.

- 'An alteration' (to an existing installation) - This box should be ticked where one or more existing circuits have been modified or extended, or items such as distribution boards and switchgear have been replaced.

Where appropriate, both the 'addition' and the 'alteration' boxes may be ticked.

THE SIGNATORIES

Two options are available for certification of the installation. Where the design, the construction and the inspection and testing are all the responsibility of one person, these three elements may be collectively certified in the section headed 'Design, Construction, Inspection and Testing'. Alternatively, where the Approved Contractor issuing the certificate has not been responsible for the design and/or the inspection and testing of the electrical work, certification of the three elements must be carried out separately using the three sections headed 'Design', 'Construction' and 'Inspection and Testing', respectively. Where one person is not responsible for all three elements, the division of responsibility should be established and agreed before commencement of the work.

The NICEIC considers that the absence of certification for the construction or the inspection and testing elements of the work would render the certificate invalid. If the 'Design' section of the certificate has not been completed, the notes on the reverse of the certificate advise the recipient to question why those responsible for the design have not certified that this important element of the work is in accordance with the national electrical safety standard.

NICEIC

This certificate is not valid if the serial number has been defaced or altered

EIC/ XXXXXXX

ELECTRICAL INSTALLATION CERTIFICATE

(REQUIREMENTS FOR ELECTRICAL INSTALLATIONS - BS 7671 [IEE WIRING REGULATIONS])

Issued by an Approved Contractor or Conforming Body enrolled with the National Inspection Council for Electrical Installation Contracting, Vintage House, 37 Albert Embankment, London SE1 7UJ.

See reverse of this page for explanatory notes relating to NICEIC software endorsement

Original (To the person ordering the work)

DETAILS OF THE CLIENT

Client / Address:

DETAILS OF THE INSTALLATION

Address:

Extent of the installation

The installation is:

New

An addition

DESIGN

I/We, being the person(s) responsible for the design of the electrical installation (as indicated by my/our signature(s) below), particulars of which are described above, having exercised reasonable skill and care when carrying out the design, hereby CERTIFY that the design work for which I/we have been responsible is, to the best of my/our knowledge and belief, in accordance with BS 7671 amended to (date) except for the departures, if any, detailed as follows:

Details of departures from BS 7671, as amended (Regulations 120-02, 120-05):

The extent of liability of the signatory/signatories is limited to the work described above as the subject of this certificate.
For the **DESIGN** of the installation: ******(*Where there is divided responsibility for the design*)

| Signature | | Date | | Name (CAPITALS) | | Designer 1 |
| Signature | | Date | | Name (CAPITALS) | ** | Designer 2 |

CONSTRUCTION

I/We, being the person(s) responsible for the construction of the electrical installation (as indicated by my/our signature below), particulars of which are described above, having exercised reasonable skill and care when carrying out the construction, hereby CERTIFY that the construction work for which I/we have been responsible is, to the best of my/our knowledge and belief, in accordance with BS 7671 amended to (date) except for the departures, if any, detailed as follows:

Details of departures from BS 7671, as amended:

The extent of liability of the signatory is limited to the work described above as the subject of this certificate.
For the **CONSTRUCTION** of the installation:

| Signature | | Date | | Name (CAPITALS) | | Constructor |

INSPECTION AND TESTING

I/We, being the person(s) responsible for the inspection and testing of the electrical installation (as indicated by my/our signatures below), particulars of which are described above, having exercised reasonable skill and care when carrying out the inspection and testing, hereby CERTIFY that the work for which I/we have been responsible is to the best of my/our knowledge and belief in accordance with BS 7671, amended to (date) except for the departures, if any, detailed as follows:

Details of departures from BS 7671, as amended:

The extent of liability of the signatory/signatories is limited to the work described above as the subject of this certificate.
For the **INSPECTION AND TESTING** of the installation: Reviewed by

| Signature | | Date | | Signature | | Date |
| Name (CAPITALS) | | Inspector | Name (CAPITALS) | | | Qualified Supervisor † |

Details of departures from BS 7671, as amended (Regulations 120-02, 120-05):

The extent of liability of the signatory is limited to the work described above as the subject of this certificate.
For the **DESIGN**, the **CONSTRUCTION** and the **INSPECTION AND TESTING** of the installation.   Reviewed by

| Signature | | Date | | Signature | | Date |
| Name (CAPITALS) | | | | Name (CAPITALS) | | Qualified Supervisor †† |

† Where the inspection and testing have been carried out by an Approved Contractor, the inspection and testing results are to be reviewed by the registered Qualified Supervisor.
†† Where the design, the construction, and the inspection and testing have been the responsibility of one person, the inspection and testing results are to be reviewed by the registered Qualified Supervisor.

Page 1 of

This form is based on the model shown in Appendix 6 of BS 7671: 1992, as amended 1997
Published by the National Inspection Council for Electrical Installation Contracting © Copyright NICEIC (Jan 2000)

Please see the 'Notes for Recipients' on the reverse of this page.

Certification for inspection and testing provides an assurance to the recipient that the results of the inspection and testing have been compared with the relevant criteria (Regulation 713-01-01) and that the electrical installation work is in accordance with *BS 7671: 1992* (as amended), except for any departures sanctioned by the designer(s) and recorded in the appropriate box(es) of the certificate.

The date of the amendment to *BS 7671* current at the time the work was carried out should be stated in the relevant space in the text contained under the headings 'Design', 'Construction' and 'Inspection and Testing' or in the section headed 'Design, Construction, Inspection and Testing', as appropriate.

Where the inspection and testing has been carried out by an Approved Contractor, the inspection and testing results should be reviewed by the registered Qualified Supervisor* who should confirm such review by signing in the appropriate one of the two boxes provided. In exceptional circumstances, the Qualified Supervisor may delegate authority (but not responsibility) for reviewing the inspection and testing results to another competent full-time electrical supervisor employed by the Approved Contractor. Only a registered Qualified Supervisor employed by the Approved Contractor is recognised by the NICEIC as eligible to sign the certificate to take responsibility for both the inspection and testing, and for reviewing the results.

DESIGN, CONSTRUCTION, INSPECTION AND TESTING

This section may be completed only where all the three elements of the installation work, namely, the design, the construction and the inspection and testing, are the responsibility of one person. Where this is not the case, the three elements should be certified separately, as indicated below.

The person signing this section of the certificate bears the same responsibility as if that person had signed the separate design, construction and inspection and testing sections of the certificate.

Where this section is utilised for collective certification of the three elements, it is not necessary to complete the three separate sections for the design, the construction, and the inspection and testing.

---

* 'Qualified Supervisor' replaces the term 'Qualifying Manager' to align with the requirements of the new industry assessment scheme expected to be introduced under the auspices of *BS EN 45011: 1998: General requirements for bodies operating product certification systems,* and with the corresponding changes in the NICEIC Rules Relating to Enrolment.

This certificate is not valid if the serial number has been defaced or altered

EIC/ **XXXXXXX**

Original (To the person ordering the work)

ELECTRICAL INSTALLATION CERTIFICATE

(REQUIREMENTS FOR ELECTRICAL INSTALLATIONS - BS 7671 [IEE WIRING REGULATIONS])

Issued by an Approved Contractor or Conforming Body enrolled with the National Inspection Council for Electrical Installation Contracting, Vintage House, 37 Albert Embankment, London SE1 7UJ.

See reverse of this page for explanatory notes relating to NICEIC software endorsement

DETAILS OF THE CLIENT

Client / Address:

DETAILS OF THE INSTALLATION

The installation is:

Address:

New

Extent of the installation covered by this certificate:

An addition

An alteration

DESIGN

I/We, being the person(s) responsible for the design of the electrical installation (as indicated by my/our signature(s) below), particulars of which are described above, having exercised reasonable skill and care when carrying out the design, hereby CERTIFY that the design work for which I/we have been responsible is, to the best of my/our knowledge and belief, in accordance with BS 7671 amended to   (date) except for the departures, if any, detailed as follows:

Details of departures from BS 7671, as amended (Regulations 120-02, 120-05):

The extent of liability of the signatory/signatories is limited to the work described above as the subject of this certificate.
For the **DESIGN** of the installation: ****** (*Where there is divided responsibility for the design*)

| Signature | | Date | | Name (CAPITALS) | | Designer 1 |
| Signature | | Date | | Name (CAPITALS) | ****** | Designer 2 |

I/we have been responsible is, to the best of my/our knowledge and belief, in accordance with BS 7671 amended to   (date) except for the departures, if any, detailed as follows:

Details of departures from BS 7671, as amended:

The extent of liability of the signatory is limited to the work described above as the subject of this certificate.
For the **CONSTRUCTION** of the installation:

| Signature | | Date | | Name (CAPITALS) | | Constructor |

INSPECTION AND TESTING

I/We, being the person(s) responsible for the inspection and testing of the electrical installation (as indicated by my/our signatures below), particulars of which are described above, having exercised reasonable skill and care when carrying out the inspection and testing, hereby CERTIFY that the work for which I/we have been responsible is to the best of my/our knowledge and belief in accordance with BS 7671, amended to   (date) except for the departures, if any, detailed as follows:

Details of departures from BS 7671, as amended:

The extent of liability of the signatory/signatories is limited to the work described above as the subject of this certificate.
For the **INSPECTION AND TESTING** of the installation: Reviewed by

| Signature | | Date | | Signature | | Date | |
| Name (CAPITALS) | | Inspector | Name (CAPITALS) | | | Qualified Supervisor † |

DESIGN, CONSTRUCTION, INSPECTION AND TESTING *

* This box to be completed only where the design, construction, inspection and testing have been the responsibility of one person.

I, being the person responsible for the design, construction, inspection and testing of the electrical installation (as indicated by my signature below), particulars of which are described above, having exercised reasonable skill and care when carrying out the design, construction, inspection and testing, hereby CERTIFY that the said work for which I have been responsible is to the best of my knowledge and belief in accordance with BS 7671, amended to   (date) except for the departures, if any, detailed as follows:

Details of departures from BS 7671, as amended (Regulations 120-02, 120-05):

The extent of liability of the signatory is limited to the work described above as the subject of this certificate.
For the **DESIGN**, the **CONSTRUCTION** and the **INSPECTION AND TESTING** of the installation.   Reviewed by

| Signature | | Date | | Signature | | Date | |
| Name (CAPITALS) | | | Name (CAPITALS) | | | Qualified Supervisor †† |

† Where the inspection and testing have been carried out by an Approved Contractor, the inspection and testing results are to be reviewed by the registered Qualified Supervisor.
†† Where the design, the construction, and the inspection and testing have been the responsibility of one person, the inspection and testing results are to be reviewed by the registered Qualified Supervisor.

Page 1 of

This form is based on the model shown in Appendix 6 of BS 7671: 1992, as amended 1997
Published by the National Inspection Council for Electrical Installation Contracting © Copyright NICEIC (Jan 2000)

Please see the 'Notes for Recipients' on the reverse of this page.

Design

Unless certification of the installation work has been undertaken by means of the section headed 'Design, Construction, Inspection and Testing', the design section should be completed by the person who has taken responsibility for the design work. A signature for this element of the certification should preferably be obtained before installation work commences, at the time the design is handed over to the installing contractor.

Where responsibility for the design is divided between the Approved Contractor and one or more other bodies, the division of responsibility should be established and agreed before commencement of the installation work. The names and signatures of the designers should be inserted in the appropriate spaces provided.

The designer is the person or organisation responsible for aspects of the installation design, such as the selection of protective devices, sizing of cables, method of installation, selection of suitable equipment, and precautions for special locations. The positioning of socket-outlets, lights etc are considered to be layout details and not electrical design, though the electrical designer will need to know the proposed positions of the equipment before completing the design calculations and determining that such positions will allow compliance with the requirements of *BS 7671*. In simple terms, the client or installation user may be able to decide where they require such items to be positioned, but the sizing and method of installing the cables and the selection of suitable equipment are matters for the designer.

Contractors that normally undertake installation work in domestic or small commercial premises only, may consider that they are not involved in installation design. However, if no other person or organisation is specifically responsible for the installation design, then the contractor will inevitably assume the role of the designer. This is the case even if the design work only involves the application of industry standards, such as for ring final circuits. In such cases the contractor will be responsible for the correct application of the industry standards and, consequently, for the whole installation design.

Details of any intended departures from *BS 7671* should be stated, but exceptionally it may be necessary to attach a separate page to the certificate rather than to attempt to give an adequate description in the box provided. In such cases, write in the box 'See details of departures on page X of this certificate'. The page number of the additional page will vary, depending on the number of schedules required.

The signatory or signatories to the design section are certifying that, with the exception of any stated departures, the design fully meets the requirements of the current issue of *BS 7671*. In signing this section, the signatory or signatories are accepting responsibility for the safety of the design.

NICEIC

This certificate is not valid if the serial number has been defaced or altered

EIC/ XXXXXXX

ELECTRICAL INSTALLATION CERTIFICATE

(REQUIREMENTS FOR ELECTRICAL INSTALLATIONS - BS 7671 [IEE WIRING REGULATIONS])

Issued by an Approved Contractor or Conforming Body enrolled with the National Inspection Council for Electrical Installation Contracting, Vintage House, 37 Albert Embankment, London SE1 7UJ.

See reverse of this page for explanatory notes relating to NICEIC software endorsement

Original (To the person ordering the work)

DETAILS OF THE CLIENT

Client / Address:

DETAILS OF THE INSTALLATION

The installation is:

Address:

Extent of the installation covered by this certificate:

New

An addition

An alteration

DESIGN

I/We, being the person(s) responsible for the design of the electrical installation (as indicated by my/our signature(s) below), particulars of which are described above, having exercised reasonable skill and care when carrying out the design, hereby CERTIFY that the design work for which I/we have been responsible is, to the best of my/our knowledge and belief, in accordance with BS 7671 amended to (date) except for the departures, if any, detailed as follows:

Details of departures from BS 7671, as amended (Regulations 120-02, 120-05):

CONSTRUCTION

I/We, being the person(s) responsible for the construction of the electrical installation (as indicated by my/our signature below), particulars of which are described above, having exercised reasonable skill and care when carrying out the construction, hereby CERTIFY that the construction work for which I/we have been responsible is, to the best of my/our knowledge and belief, in accordance with BS 7671 amended to (date) except for the the departures, if any, detailed as follows:

Details of departures from BS 7671, as amended:

The extent of liability of the signatory is limited to the work described above as the subject of this certificate.
For the **CONSTRUCTION** of the installation:

Signature	Date	Name (CAPITALS)	Constructor

INSPECTION AND TESTING

I/We, being the person(s) responsible for the inspection and testing of the electrical installation (as indicated by my/our signatures below), particulars of which are described above, having exercised reasonable skill and care when carrying out the inspection and testing, hereby CERTIFY that the work for which I/we have been responsible is to the best of my/our knowledge and belief in accordance with BS 7671, amended to (date) except for the departures, if any, detailed as follows:

Details of departures from BS 7671, as amended:

The extent of liability of the signatory/signatories is limited to the work described above as the subject of this certificate.
For the **INSPECTION AND TESTING** of the installation:

Reviewed by

Signature	Date	Signature	Date
Name (CAPITALS)	Inspector	Name (CAPITALS)	Qualified Supervisor †

DESIGN, CONSTRUCTION, INSPECTION AND TESTING *

* This box to be completed only where the design, construction, inspection and testing have been the responsibility of one person.

I, being the person responsible for the design, construction, inspection and testing of the electrical installation (as indicated by my signature below), particulars of which are described above, having exercised reasonable skill and care when carrying out the design, construction, inspection and testing, hereby CERTIFY that the said work for which I have been responsible is to the best of my knowledge and belief in accordance with BS 7671, amended to (date) except for the departures, if any, detailed as follows:

Details of departures from BS 7671, as amended (Regulations 120-02, 120-05):

The extent of liability of the signatory is limited to the work described above as the subject of this certificate.
For the **DESIGN**, the **CONSTRUCTION** and the **INSPECTION AND TESTING** of the installation.

Reviewed by

Signature	Date	Signature	Date
Name (CAPITALS)		Name (CAPITALS)	Qualified Supervisor ††

† Where the inspection and testing have been carried out by an Approved Contractor, the inspection and testing results are to be reviewed by the registered Qualified Supervisor.
†† Where the design, the construction, and the inspection and testing have been the responsibility of one person, the inspection and testing results are to be reviewed by the registered Qualified Supervisor.

Page 1 of

This form is based on the model shown in Appendix 6 of BS 7671: 1992, as amended 1997
Published by the National Inspection Council for Electrical Installation Contracting © Copyright NICEIC (Jan 2000)

Please see the 'Notes for Recipients' on the reverse of this page.

Construction

Unless certification of the installation work has been undertaken by means of the section headed 'Design, Construction, Inspection and Testing', the construction section should be completed by the person who has taken responsibility for the construction work.

Certification of the construction is usually straightforward since the certificate should be issued by the installing contractor. (In the case of an NICEIC certificate, it must be the installing Approved Contractor).

In signing this section, the signatory is certifying that the construction, except for any intended departure(s) sanctioned by the designer, satisfies the requirements of BS 7671. Such certification can only be given if the signatory has a detailed knowledge of the construction of the particular installation.

If the signatory personally carried out the installation work, this presents little difficulty. If, however, the signatory is the Qualified Supervisor or other person involved in a supervisory capacity only, the level of supervision of the installation work must have been sufficient to enable the certification to be provided with confidence.

Inspection and testing

Unless certification of the installation work has been undertaken by means of the section headed 'Design, Construction, Inspection and Testing', the inspection and testing section should be completed by the person who has taken responsibility for the inspection and testing work.

The person responsible for the inspection and testing of the new installation work should certify that the inspection and testing process has been carried out in accordance with BS 7671. If the signatory is the person who actually conducted the whole of the inspection and testing, this should be straightforward. However, if the process was conducted by more than one person, or if a Qualified Supervisor or someone in a similar supervisory capacity certifies the inspection and testing work on behalf of the Approved Contractor, that person must be entirely satisfied that the process was carried out thoroughly and in accordance with Part 7 of BS 7671, and that the results are consistent with installation work complying with all the relevant requirements of that standard.

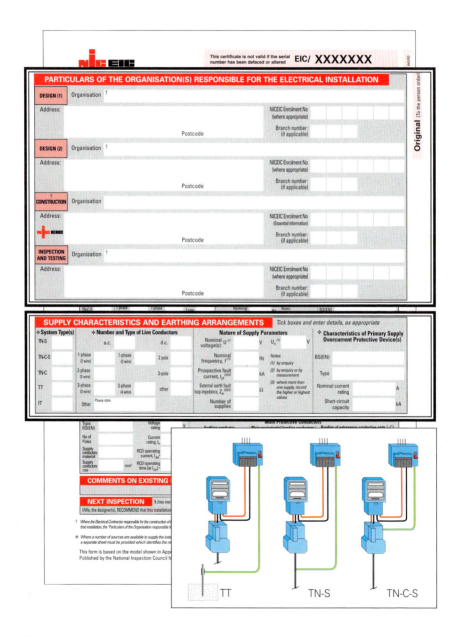

PARTICULARS OF THE ORGANISATION(S) RESPONSIBLE FOR THE ELECTRICAL INSTALLATION

Completion of the 'Particulars of the Organisation(s) Responsible for the Electrical Installation' on page 2 of the certificate will identify to the recipient the organisation(s) responsible for the work certified by their representative(s) on page 1. The organisation(s), address(es) and postcode(s) must be given, together with their NICEIC Enrolment Number and Branch Number, where appropriate. Where an Approved Contractor has been responsible for elements of the work in addition to construction, repetition of the 'Particulars of the Organisation(s) Responsible for the Electrical Installation' section is unnecessary and, in such cases, the words 'Details as given for Construction' may be inserted, where appropriate.

SUPPLY CHARACTERISTICS AND EARTHING ARRANGEMENTS

System type(s)

In the United Kingdom, it is likely that the type of system will be either TT, TN-S or TN-C-S. TN-C and IT systems are not common in the UK, and are therefore not covered in this book.

The common types of system can be briefly described by means of their particular earthing arrangements (assuming that the supply is derived from a Public Electricity Supplier (PES) at low voltage):

- **TT** - Earthing provided by the consumer's own installation earth electrode - no earthing facility is made available to the consumer by the PES, or if such a facility is made available, it is not used.

- **TN-S** - Earthing facility provided by the PES for the consumer's use - provision usually by means of a connection to the supply cable sheath or a separate protective conductor in the form of a split-concentric cable or overhead conductor.

- **TN-C-S** - Earthing facility provided by the PES, connected to the incoming supply neutral to give a Protective Multiple Earth (PME) supply, where the supply neutral and protective conductors are in the form of a Combined Neutral and Earth (CNE) conductor.

Within certain types of building or site, more than one system earthing arrangement may have been employed by the designer. For example, an installation in a sports complex may as a whole form part of a TN-C-S system, with the majority of the installation earthed to the PME earthing facility. For electrical design reasons, however, particular parts of the installation, such as a swimming pool area, may form part of a TT system. Where an installation has more than one system earthing arrangement, each arrangement needs to be recorded on the certificate.

N|C|EIC

This certificate is not valid if the serial number has been defaced or altered **EIC/ XXXXXXX**

Original (To the person ordering the work)

PARTICULARS OF THE ORGANISATION(S) RESPONSIBLE FOR THE ELECTRICAL INSTALLATION

DESIGN (1) Organisation †

Address:

Postcode

NICEIC Enrolment No (where appropriate)

Branch number: (if applicable)

DESIGN (2) Organisation †

Address:

Postcode

NICEIC Enrolment No (where appropriate)

Branch number: (if applicable)

CONSTRUCTION † Organisation

Address:

Postcode

NICEIC Enrolment No (Essential information)

Branch number: (if applicable)

INSPECTION AND TESTING Organisation †

Address:

Postcode

NICEIC Enrolment No (where appropriate)

Branch number: (if applicable)

SUPPLY CHARACTERISTICS AND EARTHING ARRANGEMENTS

Tick boxes and enter details, as appropriate

❖ System Type(s)	❖ Number and Type of Live Conductors			Nature of Supply Parameters			❖ Characteristics of Primary Supply Overcurrent Protective Device(s)	
TN-S	a.c.		d.c.	Nominal voltage(s): $U^{(1)}$	V $U_0^{(1)}$	V		
TN-C-S	1-phase (2 wire)	1-phase (3 wire)	2 pole	Nominal frequency, $f^{(1)}$	Hz	*Notes:* (1) by enquiry	BS(EN)	
TN-C	2-phase (3 wire)		3-pole	Prospective fault current, $I_{pf}^{(2)(3)}$	kA	(2) by enquiry or by measurement	Type	
TT	3-phase (3 wire)	3-phase (4 wire)	other	External earth fault loop impedance, $Z_e^{(2)(3)}$	Ω	(3) where more than one supply, record the higher or highest values	Nominal current rating	A
IT	Other *Please state*			Number of supplies			Short-circuit capacity	kA

Supplier's facility:

Installation earth electrode:

Type: (eg rods), tape etc)

Electrode resistance, R_A: (Ω)

Location:

Method of measurement:

❖ Main Switch or Circuit-Breaker
(applicable only where an RCD is suitable and is used as a main circuit-breaker)

Type: **BS(EN)**

No of Poles

Supply conductors material

Supply conductors csa mm²

Voltage rating V

Current rating, I_n A

RCD operating current, $I_{\Delta n}$ mA

RCD operating time (at $I_{\Delta n}$) ms

Maximum Demand (Load):

A per phase

Method of Protection against Indirect Contact:

Earthing conductor

Conductor material

Conductor csa mm²

Continuity check (✓)

Main Protective Conductors

Main equipotential bonding conductors		Bonding of extraneous-conductive-parts (✓)	
Conductor material		Water service	Gas service
Conductor csa	mm²	Oil service	Structural steel
Continuity check	(✓)	Lightning protection	Other incoming service(s)

COMMENTS ON EXISTING INSTALLATION

Note: Enter 'NONE' or, where appropriate, the page number(s) of additional page(s) of comments on the existing installation.

NEXT INSPECTION

§ *Enter interval in terms of years, months or weeks, as appropriate*

§

I/We, the designer(s), RECOMMEND that this installation is further inspected and tested after an interval of not more than

† *Where the Electrical Contractor responsible for the construction of the electrical installation has also been responsible for the design **and** the inspection and testing of that installation, the 'Particulars of the Organisation responsible for the Electrical Installation' may be recorded only in the section entitled 'CONSTRUCTION'.*

❖ *Where a number of sources are available to supply the installation, and where the data given for the primary source may differ from other sources, a separate sheet must be provided which identifies the relevant information relating to each additional source.*

This form is based on the model shown in Appendix 6 of BS 7671: 1992, as amended 1997
Published by the National Inspection Council for Electrical Installation Contracting © Copyright NICEIC (Jan 2000)

Page 2 of

Please see the 'Notes for Recipients' on the reverse of this page.

Number and type of live conductors

The details required to complete this section are straightforward, requiring an identification of the nature of supply current, in terms of ac or dc, and the number of the supply live conductors (including the neutral). In most cases, the supply will be single-phase (two-wire) or three-phase (three- or four-wire). Facilities for recording other configurations are also provided.

Nature of supply parameters

Provision is made in this section to record the supply parameters which comprise:

- **Nominal voltage(s)**, U (phase-to-phase), and U_O (phase-to-earth) (in Volts). This parameter can generally be determined only by enquiry. For public supplies in the UK, U is 400 V and U_O is 230 V for two-phase and three-phase supplies, and U and U_O are 230 V for single-phase supplies. Do not record measured values, however accurate.

- **Nominal frequency, f** (in Hertz). This parameter can generally be determined only by enquiry, but in the UK the nominal frequency is almost invariably 50 Hz.

- **Prospective fault current, I_{pf}** (in kA). This is the maximum fault current likely to occur in the installation, on which value the design and selection of equipment have been based. The magnitude of the prospective fault current may be obtained by 'measurement', or 'by enquiry'. Detailed guidance on the measurement of prospective fault current is given in Chapter 8. Alternatively, the designer may obtain a value of prospective fault current by enquiring at the appropriate department of the PES. Usually, neither the designer nor the person responsible for inspection and testing will have sufficient information to enable a realistic value to be calculated.

- **External earth fault loop impedance Z_e** (in Ω). As with the prospective fault current, guidance relating to the measurement of the external earth fault loop impedance (Z_e) is given in Chapter 8. The external earth fault loop impedance is measured in much the same way as the earth fault loop impedance at any other point in the installation. The alternative method, namely 'enquiry', is permitted for the determination of Z_e. If Z_e is determined by enquiry, it is normally necessary also to obtain a measured value to verify that the intended means of earthing is present and of the expected value.

- **Number of supplies**. This detail is required even where there is only one supply, in which case the entry should be '1'. Where more than one supply is available, for example the public supply and a standby generator, the total number must be recorded (in this case '2').

NICEIC

This certificate is not valid if the serial number has been defaced or altered

EIC/ XXXXXXX

PARTICULARS OF THE ORGANISATION(S) RESPONSIBLE FOR THE ELECTRICAL INSTALLATION

DESIGN (1) Organisation †

Address:

Postcode

NICEIC Enrolment No (where appropriate)

Branch number: (if applicable)

DESIGN (2) Organisation †

Address:

Postcode

NICEIC Enrolment No (where appropriate)

Branch number: (if applicable)

CONSTRUCTION † Organisation

Address:

Postcode

NICEIC Enrolment No (Essential information)

Branch number: (if applicable)

INSPECTION AND TESTING Organisation †

Address:

Postcode

NICEIC Enrolment... (where approp...

Branch nu... (if applic...

...d enter details, as appropriate

❖ **Characteristics of Primary Supply Overcurrent Protective Device(s)**

BS(EN)

Type

Nominal current rating   A

Short-circuit capacity   kA

SUPPLY CHARACTERISTICS AND EARTHING ARRANGEMENTS   Tick boxe...

❖ System Type(s)	❖ Number and Type of Live Conductors			Nature of Supply Parameters		
TN-S	a.c.		d.c.	Nominal voltage(s), $U^{(1)}$	V $U_o^{(1)}$	
TN-C-S	1-phase (2 wire)	1-phase (3 wire)	2 pole	Nominal frequency, $f^{(1)}$	Hz	Notes: (1) by enquiry
TN-C	2-phase (3 wire)		3-pole	Prospective fault current, $I_{pf}^{(2)(3)}$	kA	(2) by enquiry or by measurement
TT	3-phase (3 wire)	3-phase (4 wire)	other	External earth fault loop impedance, $Z_e^{(2)(3)}$	Ω	(3) where more than one supply, record the higher or highest values
IT	Other	Please state		Number of supplies		

Short-circuit capacity	kA

PARTICULARS OF INSTALLATION AT THE ORIGIN   Tick boxes and enter details, as appropriate

❖ **Means of Earthing**   **Details of Installation Earth Electrode (where applicable)**

Supplier's facility:

Installation earth electrode:

Type: (eg rod(s), tape etc)

Electrode resistance, R_A: (Ω)

Location:

Method of measurement:

Type: BS(EN)		Voltage rating	V	Main Protective Conductors					
				Earthing conductor	**Main equipotential bonding conductors**		**Bonding of extraneous-conductive-parts (✓)**		
No of Poles		Current rating, I_n	A	Conductor material	Conductor material		Water service	Gas service	
Supply conductors material		RCD operating current, $I_{\Delta n}$	mA	Conductor csa	mm²	Conductor csa	mm²	Oil service	Structural steel
Supply conductors csa	mm²	RCD operating time (at $I_{\Delta n}$)	ms	Continuity check (✓)	Continuity check	(✓)	Lightning protection	Other incoming service(s)

COMMENTS ON EXISTING INSTALLATION

Note: Enter 'NONE' or, where appropriate, the page number(s) of additional page(s) of comments on the existing installation.

NEXT INSPECTION § *Enter interval in terms of years, months or weeks, as appropriate*

I/We, the designer(s), RECOMMEND that this installation is further inspected and tested after an interval of not more than   §

† *Where the Electrical Contractor responsible for the construction of the electrical installation has also been responsible for the design and the inspection and testing of that installation, the 'Particulars of the Organisation responsible for the Electrical Installation' may be recorded only in the section entitled 'CONSTRUCTION'.*

❖ *Where a number of sources are available to supply the installation, and where the data given for the primary source may differ from other sources, a separate sheet may be provided which identifies the relevant information relating to each additional source.*

Page 2 of

This form is based on the model shown in Appendix 6 of BS 7671: 1992, as amended 1997
Published by the National Inspection Council for Electrical Installation Contracting © Copyright NICEIC (Jan 2000)

Please see the 'Notes for Recipients' on the reverse of this page.

Where the installation can be supplied by more than one source, such as the public supply and a standby generator, the higher or highest values of prospective fault current, I_{pf}, and external earth fault loop impedance, Z_e, must be recorded in the data-entry boxes provided for this purpose.

Characteristics of primary supply overcurrent protective device(s)

The information to be recorded here is the British Standard (or other appropriate Standard) product specification in terms of BS (EN) number, together with the type, the nominal current rating (I_n), and the short-circuit capacity of the device. This information should be obtained from the installation designer at an early stage and confirmed by inspection of the markings on the PES's cut-out fuse-holder. If no clear indication is given on the overcurrent device, confirmation of its type and rating should be sought from the PES.

Where a number of sources are available to supply the installation, and where the data given for the primary source may differ from other supplies, an additional page should be included in the certificate giving the relevant information for each additional supply.

PARTICULARS OF INSTALLATION AT THE ORIGIN

Means of earthing

Where protection against indirect contact is provided by Earthed Equipotential Bonding and Automatic Disconnection of supply (EEBAD), there is a need to provide a connection of all exposed-conductive-parts and extraneous-conductive-parts to a means of earthing via the main earthing terminal. The information required here is for the purpose of identifying the particular means of earthing used, in terms of a **supplier's facility** and/or an **installation earth electrode**.

Details of installation earth electrode

Every TT system must have an installation earth electrode. If the system is not TT and there is no earth electrode, write 'NONE' for the type and 'N/A' (Not Applicable) for the location. Do not leave any boxes empty.

If the installation has an earth electrode, give a brief description of its type. The types of earth electrode recognised by *BS 7671* are listed in Regulation 542-02-01. Metal pipework forming part of a gas, water or any other service network must not be used as an earth electrode.

Sufficient details must be given under 'location' so that persons unfamiliar with the installation and building layout will be able to locate the electrode for periodic inspection and testing purposes.

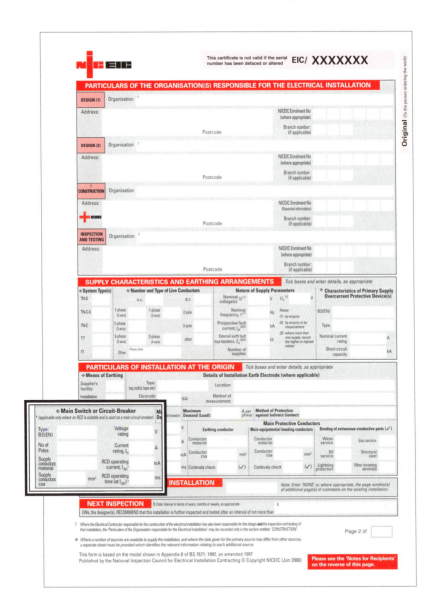

The earth electrode resistance, R_A, must be measured and the ohmic value recorded in the space provided for the purpose.

The method used to measure the earth electrode resistance to earth must also be recorded. Guidance on the testing procedure is given in Chapter 8. Two methods of measurement are recognised by *IEE Guidance Note 3: Inspection and Testing:*

- Using a proprietary earth electrode test instrument.
- Using an earth fault loop impedance test instrument.

Where suitable and sufficient ground area is available for measurement purposes, a proprietary earth electrode test instrument may be used. Where hard surfaces or lack of space make the use of this method impracticable, the use of an earth fault loop impedance test instrument is appropriate.

Factors to be considered when choosing the method to use are given in Chapter 8.

Main switch or circuit-breaker

Type BS (EN)

This is normally a straightforward process of copying down the details from the installation design, having checked that the installed device conforms to that design. The types of main switch and circuit-breaker commonly used should have the relevant British Standard or BS EN number clearly marked. If, after checking both the design information and the device, no BS number or any other meaningful identification of the type of the device can be found, the lack of this essential information should be referred to the designer for clarification. Do not leave the box blank.

Voltage rating

This is normally a straightforward process of copying down the relevant details from the installation design, having first checked that the device has been selected and installed in accordance with that design.

Number of poles

The number of poles is usually 2 for a single-phase device, and 3 or 4 for a three-phase device. The actual number of poles must be confirmed before writing the number in the box, as a variety of switching arrangements are available.

Current rating (I_n)

The information required here is the current rating, which is normally clearly indicated on the device (or on the fuse, if provided). This should be checked against the design information.

NICEIC

This certificate is not valid if the serial number has been defaced or altered **EIC/ XXXXXXX**

Original (To the person ordering the work)

PARTICULARS OF THE ORGANISATION(S) RESPONSIBLE FOR THE ELECTRICAL INSTALLATION

DESIGN (1) Organisation †
Address:
Postcode
NICEIC Enrolment No (where appropriate)
Branch number: (if applicable)

DESIGN (2) Organisation †
Address:
Postcode
NICEIC Enrolment No (where appropriate)
Branch number: (if applicable)

† CONSTRUCTION Organisation
Address:
Postcode
NICEIC Enrolment No (Essential information)
Branch number: (if applicable)

INSPECTION AND TESTING Organisation †
Address:
Postcode
NICEIC Enrolment No (where appropriate)
Branch number: (if applicable)

SUPPLY CHARACTERISTICS AND EARTHING ARRANGEMENTS
Tick boxes and enter details, as appropriate

❖ System Type(s)	❖ Number and Type of Live Conductors		Nature of Supply Parameters	❖ Characteristics of Primary Supply Overcurrent Protective Device(s)
TN-S	a.c.	d.c.	Nominal voltage(s), $U^{(1)}$ V $U_0^{(1)}$ V	
TN-C-S	1-phase (2 wire)	1-phase (3 wire)	2 pole	Nominal frequency, $f^{(1)}$ Hz	BS(EN)
TN-C	2-phase (3 wire)		3-pole	Prospective fault current, $I_{pf}^{(2)(3)}$ kA	Type
TT	3-phase (3 wire)	3-phase (4 wire)	other	External earth fault loop impedance, $Z_e^{(2)(3)}$ Ω	Nominal current rating A
IT	Other Please state			Number of supplies	Short-circuit capacity kA

Notes:
(1) by enquiry
(2) by enquiry or by measurement
(3) where more than one supply, record the higher or highest values

PARTICULARS OF INSTALLATION AT THE ORIGIN
Tick boxes and enter details, as appropriate

❖ Means of Earthing		Details of Installation Earth Electrode (where applicable)
Supplier's facility:	Type: (eg rod(s), tape etc)	Location:
Installation	Electrode (Ω)	Method of

❖ Main Switch or Circuit-Breaker
* (applicable only where an RCD is suitable used as a main circuit-breaker)

		Maximum Demand (Load):	A per phase	Method of Protection against Indirect Contact:

Type: BS(EN)		Voltage rating	V	**Main Protective Conductors**				
No of Poles		Current rating, I_n	A	Earthing conductor		Main equipotential bonding conductors		Bonding of extraneous-conductive-parts (✓)
Supply conductors: material		RCD operating current, $I_{Δn}$*	mA	Conductor material		Conductor material		Water service / Gas service
Supply conductors: csa	mm²	RCD operating time (at $I_{Δn}$)*	ms	Conductor csa mm²		Conductor csa mm²		Oil service / Structural steel
				Continuity check (✓)		Continuity check (✓)		Lightning protection / Other incoming service(s)

NEXT INSPECTION
§ Enter interval in terms of years, months or weeks, as appropriate

(I/We, the designer(s), RECOMMEND that this installation is further inspected and tested after an interval of not more than _____ §

† Where the Electrical Contractor responsible for the construction of the electrical installation has also been responsible for the design **and** the inspection and testing of that installation, the 'Particulars of the Organisation responsible for the Electrical Installation' may be recorded only in the section entitled 'CONSTRUCTION'.

❖ Where a number of sources are available to supply the installation, and where the data given for the primary source may differ from other sources, a separate sheet must be provided which identifies the relevant information relating to each additional source.

This form is based on the model shown in Appendix 6 of BS 7671: 1992, as amended 1997
Published by the National Inspection Council for Electrical Installation Contracting © Copyright NICEIC (Jan 2000)

Page 2 of _____

Please see the 'Notes for Recipients' on the reverse of this page.

Supply conductors: material

The supply conductor material is required to be identified here. This is generally the conductor material of the 'tails' from the PES's meter to the consumer's installation main switch, usually copper.

Supply conductors: csa

This is the cross-sectional area of the supply conductors (for example, 50 mm^2).

Where the main circuit-breaker is a residual current device

The BS (EN) number, voltage rating, number of poles and current rating should be determined by examining both the design information and the installed device. The rating referred to in this case is the rated current, I_n, and **not** the rated residual operating (tripping) current, $I_{\Delta n}$, of the device.

RCD operating current ($I_{\Delta n}$)

This should be determined from the design information and checked with the marking on the device. The rated residual operating (tripping) current will usually be given in units of mA (for example '300 mA'), but in some instances it may be stated in amperes (for example '0.3 A'). An RCD may operate (trip) at a current just a little more than half its rated residual operating current. Where an RCD is used as a main circuit-breaker, it should generally not have a low rated residual operating current (ie below 100 mA), in order to minimise the risk of unwanted tripping.

RCD operating time (at $I_{\Delta n}$)

Guidance on this test, which requires the use of an RCD test instrument, is given in Chapter 8. Record in the box the measured operating time in milliseconds when subjected to a test current equal to the rated residual operating current, $I_{\Delta n}$.

Maximum demand (load)

This is not, as is sometimes assumed, the rating of the PES's cut-out fuse(s). The maximum demand is a value, expressed in amperes per phase, evaluated on the basis of the connected load with an allowance for diversity. The installation designer would normally be expected to provide the maximum demand value. For smaller installations, guidance on estimating the maximum demand is given in *IEE Guidance Note 1: Selection and Erection*. There are, however, other methods by which the maximum demand current may be assessed, and these are not precluded provided they give realistic values.

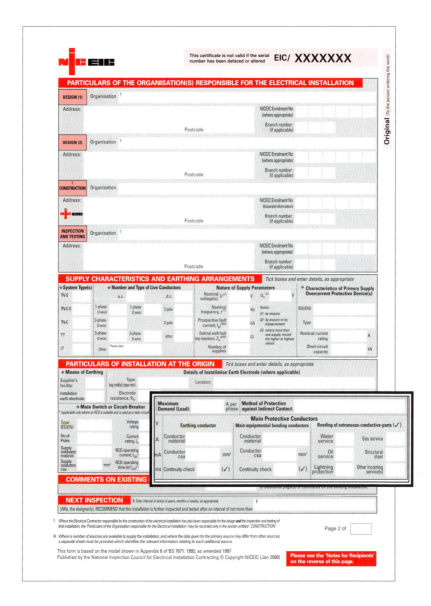

In the case of alterations and additions, it is essential that an assessment of the maximum demand of the existing installation is made prior to carrying out modifications. To this assessment of the existing maximum demand must be added all additional loads resulting from the proposed new work, taking account of diversity as appropriate, so that the suitability of the incoming electricity service to supply the proposed increase in load can be ascertained.

Method of protection against indirect contact

This refers to the principal method of protection against indirect contact. In many, if not most, cases the method will be Earthed Equipotential Bonding and Automatic Disconnection of supply (EEBAD). The particular method the designer has chosen to use from those listed in Regulation 413-01-01 must be fully described here.

Main protective conductors

Earthing conductor

The earthing conductor is the protective conductor that connects the main earthing terminal of the installation to the means of earthing. The conductor material and cross-sectional area of the earthing conductor must be stated here, followed by a tick to confirm that its continuity has been checked. This indication confirms that the inspector has tested the earthing conductor, and/or inspected it throughout its length and checked its connections.

Main equipotential bonding conductors

The conductor material and cross-sectional area of the main bonding conductors must be stated here, followed by a tick to confirm that their continuity has been checked. This indication confirms that the inspector has tested the continuity of the bonding conductors and/or inspected them throughout their length and checked their connections.

Bonding of extraneous-conductive-parts

Provision is made to record the extraneous-conductive-parts to which main bonding to the main earthing terminal of the installation has been effected, such as the incoming water and gas services.

Inspection, Testing, Certification and Reporting

Original [To the person ordering the work]

PARTICULARS OF THE ORGANISATION(S) RESPONSIBLE FOR THE ELECTRICAL INSTALLATION

DESIGN (1) Organisation †

Address:

NICEIC Enrolment No (where appropriate)

Postcode

Branch number: (if applicable)

DESIGN (2) Organisation †

Address:

NICEIC Enrolment No (where appropriate)

Postcode

Branch number: (if applicable)

† CONSTRUCTION Organisation

Address:

NICEIC Enrolment No (Essential information)

Postcode

Branch number: (if applicable)

INSPECTION AND TESTING Organisation †

Address:

NICEIC Enrolment No (where appropriate)

Postcode

Branch number: (if applicable)

SUPPLY CHARACTERISTICS AND EARTHING ARRANGEMENTS
Tick boxes and enter details, as appropriate

❖ System Type(s)	❖ Number and Type of Live Conductors			Nature of Supply Parameters			❖ Characteristics of Primary Supply Overcurrent Protective Device(s)	
TN-S		a.c.	d.c.	Nominal voltage(s): $U_0^{(1)}$	V $U_n^{(1)}$	V	
TN-C-S	1-phase (2 wire)	1-phase (3 wire)	2 pole	Nominal frequency, $f^{(1)}$	Hz	Notes: (1) by enquiry	BS(EN)
TN-C	2-phase (3 wire)		3-pole	Prospective fault current, $I_{pf}^{(2)(3)}$	kA	(2) by enquiry or by measurement	Type
TT	3-phase (3 wire)	3-phase (4 wire)	other	External earth fault loop impedance, $Z_e^{(2)(3)}$	Ω	(3) where more than one supply, record the higher or highest values	Nominal current rating	A
IT	Other Please state			Number of supplies			Short-circuit capacity	kA

PARTICULARS OF INSTALLATION AT THE ORIGIN
Tick boxes and enter details, as appropriate

❖ Means of Earthing		Details of Installation Earth Electrode (where applicable)	
Supplier's facility:	Type: (eg rods), tape etc)	Location:	
Installation earth electrode:	Electrode resistance, R_A:	(Ω)	Method of measurement:

❖ Main Switch or Circuit-Breaker		Maximum Demand (Load):	A per phase	Method of Protection against Indirect Contact:	
(applicable only where an RCD is suitable and is used as a main circuit-breaker)					
Type: BS(EN)	Voltage rating	V		**Main Protective Conductors**	
			Earthing conductor	**Main equipotential bonding conductors**	**Bonding of extraneous-conductive-parts (✓)**
No of Poles	Current rating, I_n	A	Conductor material	Conductor material	Water service	Gas service
Supply conductors material	RCD operating current, $I_{\Delta n}$	mA	Conductor csa	Conductor csa	mm²	Oil service	Structural steel
Supply conductors	mm²	RCD operating	ms	Continuity check	(✓)	Continuity check	(✓)	Lightning	Other incoming

COMMENTS ON EXISTING INSTALLATION

Note: Enter 'NONE' or, where appropriate, the page number(s) of additional page(s) of comments on the existing installation.

NEXT INSPECTION

§ *Enter interval in terms of years, months or weeks, as appropriate* §

I/We, the designer(s), RECOMMEND that this installation is further inspected and tested after an interval of not more than

---

† Where the Electrical Contractor responsible for the construction of the electrical installation has also been responsible for the design **and** the inspection and testing of that installation, the 'Particulars of the Organisation responsible for the Electrical Installation' may be recorded only in the section entitled 'CONSTRUCTION'.

❖ Where a number of sources are available to supply the installation, and where the data given for the primary source may differ from other sources, a separate sheet must be provided which identifies the relevant information relating to each additional source.

This form is based on the model shown in Appendix 6 of BS 7671: 1992, as amended 1997
Published by the National Inspection Council for Electrical Installation Contracting © Copyright NICEIC (Jan 2000)

Page 2 of

Please see the 'Notes for Recipients' on the reverse of this page.

COMMENTS ON EXISTING INSTALLATION

This section of the certificate is applicable only where the work carried out is an addition or alteration to an existing installation, or where a complete rewire of an installation has been carried out and the installation is connected to an existing main incoming supply. Any deficiencies observed in the existing installation which do not affect the safety of the new work must be recorded here, by reference to an additional sheet of the certificate giving the detailed comments. Examples of such deficiencies are deterioration and minor defects. Where there are no comments to be made, the entry should read 'None'.

It is important to appreciate that any defects which would result in a reduced level of safety in the new work (that is a level of safety less than would be afforded by compliance with *BS 7671*), must be corrected before the new work is put into service. Consequently, no such defects should be recorded on the Electrical Installation Certificate.

In particular, by the time the installation work is ready to be certified and put into service, the contractor should have ensured that the rating and condition of any existing equipment, such as the main incoming supply, cables and switchgear, which will have to carry any additional load as a result of the new work, are adequate for the altered circumstances and that the existing earthing arrangement, main equipotential bonding and, where appropriate, supplementary equipotential bonding are all adequate.

NEXT INSPECTION

This section should be completed by the person responsible for the design of the installation work. A time interval, in terms of years, months or weeks, should be inserted to indicate when the next inspection will be due. *IEE Guidance Note 3: Inspection and Testing* gives guidance in terms of the maximum intervals between initial certification and the first periodic inspection and testing for various types of premises. The interval recorded on the certificate should take account of the available guidance material, any mandatory inspection requirements relating to the particular installation, and any other special circumstances that prevail. Under no circumstances should 'N/A' (or 'Not Applicable') be inserted.

Inspection, Testing, Certification and Reporting

SCHEDULES

The remainder of the Electrical Installation Certificate comprises various schedules. Guidance on how to complete these schedules is given in Chapter 6 ('Schedules'). Guidance on the process of inspection and testing necessary to complete the schedules will be found in Chapter 7 ('Inspection') and Chapter 8 ('Testing').

Where required, continuation schedules are available separately from the NICEIC. Additional (continuation) schedules are required where the installation includes distribution ('sub-main') circuits - see Chapter 6.

The total number of pages comprising the certificate must be inserted in the box provided at the foot of each of the pages on the right-hand side.

Chapter 3

The Domestic Electrical Installation Certificate

Inspection, Testing, Certification and Reporting

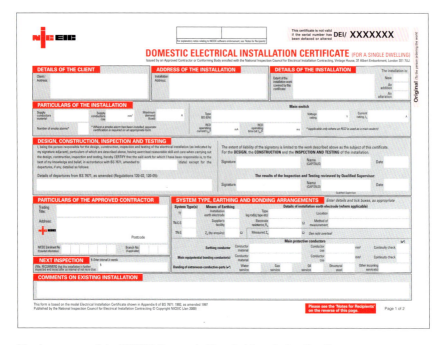

The front page of the NICIEC Domestic Electrical Installation Certificate.

Subject to certain limitations, the certificate is designed to be used for electrical installation work relating to a single dwelling (house or individual flat).

3 **THE DOMESTIC ELECTRICAL INSTALLATION CERTIFICATE**

REQUIREMENTS AND LIMITATIONS

BS 7671 requires a standard form of Electrical Installation Certificate (including a schedule of test results) to be issued for all new installation work, and for all alterations and additions to existing installations. A separate certificate must be issued for each distinct electrical installation. The certificate provides a formal assurance that the installation complies with the requirements of the national standard for electrical safety.

A certificate (based on the model form given in *BS 7671*) must be issued to the person who ordered the work, whether or not the contractor is registered with the NICEIC, and whether or not a certificate has been specifically requested by the client. The NICEIC strongly prefers Approved Contractors to issue NICEIC certificates, as this provides a measure of confidence to the recipients. Non-approved contractors are not authorised to issue NICEIC certificates, but similar certificates based on the model form given in *BS 7671* are available from the NICEIC.

It should be noted that the Domestic Electrical Installation Certificate is not intended to be issued to confirm the completion of a contract. The certificate is a declaration of electrical safety which should be issued before an installation is put into service. The certificate should not be withheld for contractual reasons, especially if the electrical installation is available for use.

The Domestic Electrical Installation Certificate is to be used only for the initial certification of a new installation, or of new work associated with an alteration or addition to an existing installation, carried out in accordance with *BS 7671: 1992* as amended. To contain the certificate within two pages, it has been necessary to impose particular limitations on its use. The certificate may therefore only be used where **all** the following conditions apply:

- The electrical installation work relates to a single dwelling (house or individual flat).

- The design, the construction, and the inspection and testing of the electrical installation work has been the responsibility of one person.

- The supply to the installation is single-phase, 50 Hz, and the nominal voltage does not exceed 230 V.

- The maximum demand for the installation, with due allowance for diversity, does not exceed 100 A.

- The installation forms part of a TT, TN-S or TN-C-S (PME) system.

- The supplier's cut-out incorporates an HBC fuse to BS 88, or a BS 1361 Type II fuse, rated at 100 A or less.

- The supply fault level does not exceed 16 kA.

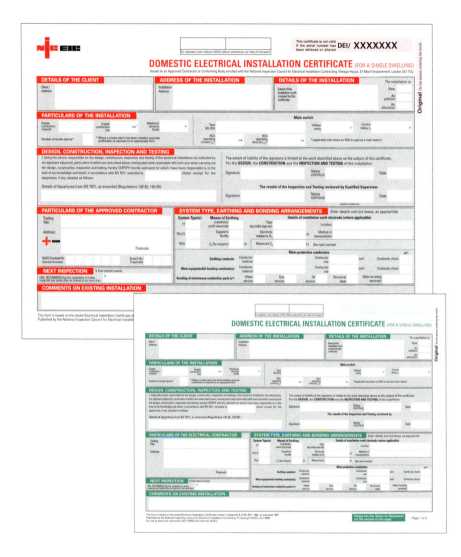

Chapter 3

- Protection against indirect contact is provided primarily by Earthed Equipotential Bonding and Automatic Disconnection of supply (EEBAD).

- The entire installation has not more than one consumer unit, located at the origin of the installation (i.e. no distribution [sub-main] circuits).

- The consumer unit does not provide for more than twelve final circuits.

- The installation is supplied from one source only.

In the case of new installation work associated with an alteration or an addition to an existing installation, the installation when modified must meet all of the above conditions if the Domestic Electrical Installation Certificate is to be used. Otherwise, the Electrical Installation Certificate must be used.

For all other installation work in premises, including work in common areas of apartment blocks etc, and where the Electricity at Work Regulations apply under normal circumstances (ie places where a work activity is undertaken), an Electrical Installation Certificate (see Chapter 2) should be used, unless the work falls within the limited scope of the Minor Electrical Installation Works Certificate (see Chapter 4).

Where a certificate is to be issued for an alteration or addition to an existing installation, the designer is required to ascertain that the rating and condition of any existing equipment, including that of the supplier (which may have to carry any additional load), are adequate to accommodate in safety the altered circumstances resulting from the modifications, and that the earthing and bonding arrangements are also adequate (see Regulation 130-09-01).

Electrical contractors that undertake installation work in domestic premises may consider that they are not involved in installation design. However, for installation work falling within the scope of the Domestic Electrical Installation Certificate, the contractor will invariably have assumed the role of the designer, even where the design work only involves the application of industry standards, such as for ring final circuits. In such cases, the contractor is responsible for the correct application of the industry standards and, consequently, for the design of the installation work.

The design of electrical installation work includes such aspects as the selection of protective devices, sizing of cables, methods of installation, selection of suitable equipment, and precautions for special locations such as bathrooms. The positioning of socket-outlets, lights etc are considered to be layout details and not electrical design, though the electrical designer will need to know the proposed positions of the equipment before completing the design calculations and determining that such positions will allow compliance with the requirements of *BS 7671*. In simple terms, the client or installation user may be able to decide where they require such items to be

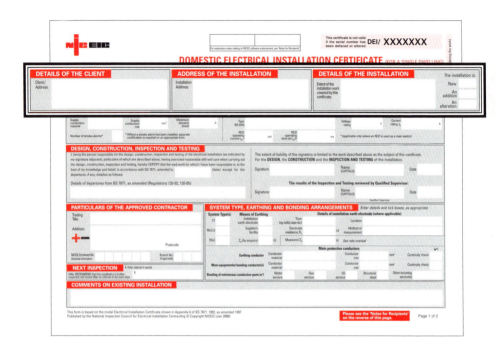

positioned, but the sizing and method of installing the cables and the selection of suitable equipment are matters for the designer of the installation work.

Where an electrical contractor discovers the existence of a dangerous or potentially dangerous situation in the existing installation (such as the absence of earthing or main bonding where the method of protection against indirect contact is EEBAD), the new work should not proceed and the client should be advised immediately, preferably in writing, to satisfy the duties imposed on competent persons by the *Electricity at Work Regulations 1989.*

An NICEIC Domestic Electrical Installation Certificate may be issued only by the Approved Contractor responsible for the installation work. A certificate must never be issued to cover another contractor's installation work.

An NICEIC Domestic Electrical Installation Certificate consists of two pages, both of which must be fully completed and issued. Generally, all unshaded data-entry boxes should be completed by inserting the necessary text, a 'Yes' or a '✔' to indicate that the task has been completed, or a numeric value of the measured parameter, or exceptionally by entering 'N/A' meaning 'Not Applicable', where appropriate.

Regardless of the method of compilation of the certificate (software-assisted or by hand), it remains the responsibility of the compiler of the certificate to ensure that the information provided on the certificate is factual, and that the electrical installation work to which the certificate relates is safe to be put into service.

Where the electrical installation work to which the certificate relates includes the provision of a mains-powered fire detection and alarm system (such as one or more smoke alarms), the Domestic Electrical Installation Certificate must be accompanied by a separate certificate in accordance with British Standard *BS 5839: Part 6: Code of Practice for the design and installation of fire detection and alarm systems in dwellings.*

COMPILATION OF THE CERTIFICATE

The remainder of this Chapter provides guidance on completing each section of the certificate.

DETAILS OF THE CLIENT

The details to be provided are the name and address of the person or organisation that engaged the contractor to carry out the installation work. Where, for example, the electrical contractor is engaged by a builder, the details entered should be those of the builder. In the case of lengthy name and address details, suitable abbreviations may be used if acceptable to the client.

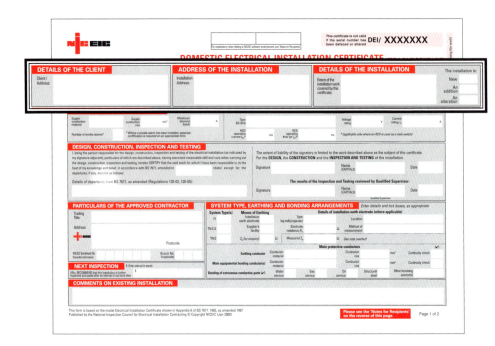

ADDRESS OF THE INSTALLATION

The address to be provided is the complete postal address of the installation, including the postcode, where known.

DETAILS OF THE INSTALLATION

Extent of the installation work covered by this certificate

For a new installation, the certificate may be intended to apply to the whole of the electrical installation at the address given. If not, it is essential to clearly define the extent of the installation to which the certificate applies in terms of areas or services; for example, 'whole electrical installation in refurbished kitchen'.

The description will depend on the particular circumstances, and must be tailored to identify explicitly the extent of the installation work covered by the certificate. In the case of additions or alterations, particular care must be taken to define exactly what work the certificate covers.

The box should be used to identify and clarify any particular exclusions agreed with the client. Any other exclusions that may not be readily apparent to the client or the installation user should also be recorded here.

It should be appreciated that failure to clearly describe the extent of the work covered by the certificate could involve the contractor in unforeseen liabilities at a later date.

Nature of the installation work

Three boxes are provided to enable the contractor to identify the nature of the installation work:

- 'New' - This box should be ticked only if the whole installation has been installed as new, or if a complete rewire has been carried out.
- 'An addition' (to an existing installation) - This box should be ticked if an existing installation has been modified by the addition of one or more new circuits.
- 'An alteration' (to an existing installation) - This box should be ticked where one or more existing circuits have been modified or extended, or where, for example, the consumer unit has been replaced.

Where appropriate, both the 'addition' and the 'alteration' boxes may be ticked.

Inspection, Testing, Certification and Reporting

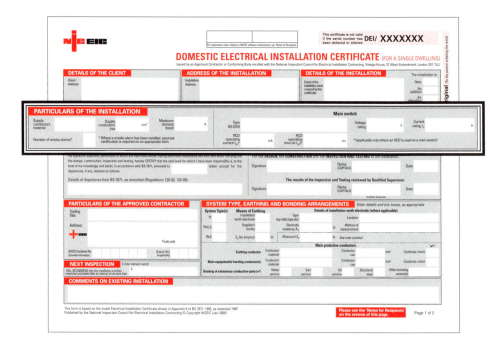

60

PARTICULARS OF THE INSTALLATION

As the use of the Domestic Electrical Installation Certificate is restricted to installation work where the design, the construction, and the inspection and testing of the electrical installation work has been the responsibility of one person, the following information should be readily available from that person's installation design:

Supply conductors: material

The supply conductor material is required to be identified here. This is generally the conductor material of the 'tails' from the Public Electricity Supplier's (PES) meter to the consumer's installation main switch, usually copper.

Supply conductors: csa

This is the cross-sectional area of the supply conductors (tails), for example, 25 mm^2.

Maximum demand (load)

This is not, as is sometimes assumed, the rating of the PES's cut-out fuse. The maximum demand is a value, expressed in amperes, evaluated on the basis of the connected load with an allowance for diversity. The value would be determined at the design stage. Guidance on estimating maximum demand is given in *IEE Guidance Note 1: Selection and Erection*. There are, however, other methods by which the maximum demand current may be assessed, and these are not precluded provided they give realistic values.

In the case of alterations and additions, it is essential that an assessment of the maximum demand of the existing installation is made prior to carrying out modifications. To this assessment of the existing maximum demand must be added all additional loads resulting from the proposed new work, taking account of diversity as appropriate, so that the suitability of the incoming electricity service to supply the proposed increase in load can be ascertained.

Number of smoke alarms

Where the electrical work involves the installation of one or more smoke alarms, record the number of smoke alarms that have been installed. It should be noted that the installation of smoke alarms must be separately certificated in accordance with *BS 5839: Part 6: Code of practice for the design and installation of fire detection and alarm systems in dwellings*.

Inspection, Testing, Certification and Reporting

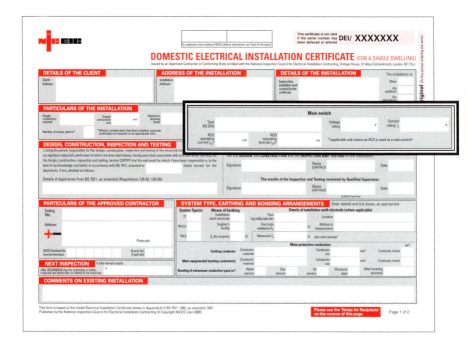

Main switch

Type BS (EN)

This is normally a straightforward process of copying down the details from the previously completed installation design, having checked that the installed switch conforms to that design. The types of main switch commonly used in domestic premises should have the relevant British Standard or BS EN number clearly marked on them.

Voltage rating

This is normally a straightforward process of copying down the relevant details from the installation design, having first checked that the installed switch conforms to that design.

Current rating (I_n)

The information required here is the current rating, which should be clearly indicated on the switch. This should be checked against the design information.

Where an RCD is used as the main switch

The BS (EN) number, voltage rating and current rating should be determined by examining both the design information and the installed switch. The rating referred to in this case is the rated current, I_n, and **not** the rated residual operating (tripping) current, $I_{\Delta n}$, of the device.

RCD operating current ($I_{\Delta n}$)

This should be determined from the design information and checked with the marking on the switch. The rated residual operating (tripping) current will usually be given in units of mA (for example '100 mA'), but in some instances it may be stated in amperes (for example '0.1 A'). An RCD may operate (trip) at a current just a little more than half its rated residual operating current. Where an RCD is used as the main switch, it should generally not have a low rated residual operating current (ie below 100 mA), in order to minimise the risk of unwanted tripping.

RCD operating time (at $I_{\Delta n}$)

Guidance on this test, which requires the use of an RCD test instrument, is given in Chapter 8. Record in the box the measured operating time in milliseconds when subjected to a test current equal to the rated residual operating current, $I_{\Delta n}$.

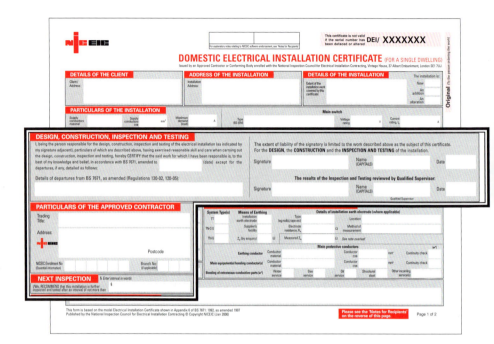

DOMESTIC ELECTRICAL INSTALLATION CERTIFICATE (FOR A SINGLE DWELLING)

Issued by an Approved Contractor or Conforming Body enrolled with the National Inspection Council for Electrical Installation Contracting, Vintage House, 37 Albert Embankment, London SE1 7UJ

This certificate is not valid if the serial number has been defaced or altered DEI/ XXXXXXX

DETAILS OF THE CLIENT

ADDRESS OF THE INSTALLATION

DETAILS OF THE INSTALLATION

PARTICULARS OF THE INSTALLATION

DESIGN, CONSTRUCTION, INSPECTION AND TESTING

PARTICULARS OF THE APPROVED CONTRACTOR

NEXT INSPECTION

This form is based on the model Electrical Installation Certificate shown in Appendix 6 of BS 7671: 1992, as amended 1997
Published by the National Inspection Council for Electrical Installation Contracting © Copyright NICEIC (Jan 2000)

Please see the 'Notes for Recipients' on the reverse of this page.

Page 1 of 2

DESIGN, CONSTRUCTION, INSPECTION AND TESTING

Completion of this section is essential as it provides for certification of the three elements of the installation work; the design, the construction, and the inspection and testing. In signing this section, the signatory is accepting responsibility for the safety of the completed installation work, including its design and construction.

The signatory is also certifying that, with the exception of any departures recorded on the certificate, the design, the construction, and the inspection and testing fully meet the requirements of the current issue of *BS 7671*. The signatory must be entirely satisfied that the process of inspection and testing was carried out thoroughly and in accordance with Part 7 of *BS 7671*, and that the results are consistent with installation work complying with all the relevant requirements of that standard.

The date of the amendment to *BS 7671* current at the time the electrical work was carried out should be stated in the relevant space in the text. Details of any departures from *BS 7671* should be entered in the space provided.

The inspection and testing results should be reviewed by the registered Qualified Supervisor* who should confirm such review by signing in the appropriate space provided. By signing this section, the Qualified Supervisor is confirming that the certificate has been completed satisfactorily prior to issue.

Only a registered Qualified Supervisor employed by the Approved Contractor is recognised by the NICEIC as eligible to sign the certificate to take responsibility for reviewing the results of the inspection and testing.

PARTICULARS OF THE APPROVED CONTRACTOR

Completion of the 'Particulars of the Approved Contractor' will identify to the recipient the contractor responsible for the work which is the subject of the certificate. The Approved Contractor's trading title, address and postcode must be given, together with its NICEIC Enrolment Number and Branch Number, where appropriate.

---

* 'Qualified Supervisor' replaces the term 'Qualifying Manager' to align with the requirements of the new industry assessment scheme expected to be introduced under the auspices of *BS EN 45011: 1998: General requirements for bodies operating product certification systems,* and with corresponding changes in the NICEIC Rules Relating to Enrolment.

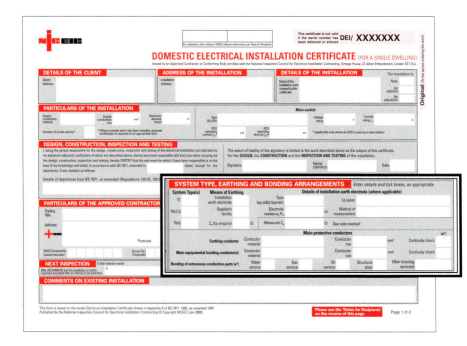

SYSTEM TYPE, EARTHING AND BONDING ARRANGEMENTS

System type(s)

In the United Kingdom, it is likely that, for domestic accommodation, the system type will be either TT, TN-S or TN-C-S, each of which falls within the scope of the Domestic Electrical Installation Certificate.

The types of system can be briefly described by means of their particular earthing arrangements:

- **TT** - Earthing provided by the consumer's own installation earth electrode - no earthing facility is made available to the consumer by the PES or, if such a facility is made available, it is not used.

- **TN-S** - Earthing facility provided by the PES for the consumer's use - provision usually by means of a connection to the supply cable sheath or a separate protective conductor in the form of a split-concentric cable or overhead conductor.

- **TN-C-S** - Earthing facility provided by the PES, connected to the incoming supply neutral to give a Protective Multiple Earth (PME) supply, where the supply neutral and protective conductors are in the form of a Combined Neutral and Earth (CNE) conductor.

For some dwellings, more than one system earthing arrangement may be used. For example, the installation within a dwelling may form part of a TN-C-S system, being earthed to the PME earthing facility. However, another part of the installation, such as a final circuit supplying a garage or shed, may form part of a TT system. Where an installation has two system earthing arrangements (one of which must essentially be a TT system), both types are required to be recorded on the certificate (ie TN-S or TN-C-S, and TT).

Means of earthing

The protection against indirect contact provided by Earthed Equipotential Bonding and Automatic Disconnection of supply (EEBAD) requires the connection of all exposed-conductive-parts and extraneous-conductive-parts to a means of earthing via the main earthing terminal. The information required here is for the purpose of identifying the particular means of earthing provided, in terms of a **supplier's facility** and/or an **installation earth electrode**.

Details of installation earth electrode

Every TT system must have an installation earth electrode. If the system is not TT and there is no earth electrode, write 'NONE' for the type and 'N/A' (Not Applicable) for the location. Do not leave any boxes empty.

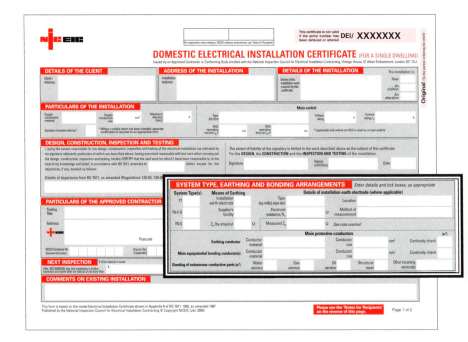

If the installation has an earth electrode, give a brief description of its type. The types of earth electrode recognised by *BS 7671* are listed in Regulation 542-02-01. Metal pipework forming part of a gas, water or any other service network must not be used as an earth electrode.

Sufficient details must be given under 'location' so that persons unfamiliar with the installation and building layout will be able to locate the electrode for periodic inspection and testing purposes.

The earth electrode resistance, R_A, must be measured and the ohmic value recorded in the space provided for the purpose.

The method used to measure the earth electrode resistance to earth must also be recorded. Guidance on the test procedure is given in Chapter 8. Two methods of measurement are recognised by *IEE Guidance Note 3: Inspection and Testing:*

- Using a proprietary earth electrode test instrument.

- Using an earth fault loop impedance test instrument.

Where suitable and sufficient ground area is available for measurement purposes, a proprietary earth electrode test instrument may be used. Where hard surfaces or lack of space make the use of this method impracticable, the use of an earth fault loop impedance test instrument is appropriate.

Factors to be considered when choosing the test method to use are given in Chapter 8.

Supplier's facility

Where the Public Electricity Supplier provides an earthing facility, and it has been used to earth the installation, this must be indicated in the space provided.

Z_e (external earth fault loop impedance) by enquiry

Where the means of earthing is the supplier's facility, the external earth fault loop impedance, Z_e, should be determined by enquiry to the supplier, for design purposes. The value of Z_e obtained by enquiry and should be recorded on the data-entry box provided for this purpose.

Measured Z_e (external earth fault loop impedance)

Even where a value of external earth fault loop impedance has been determined by enquiry, it is normally necessary to obtain a measured value of Z_e to verify that the

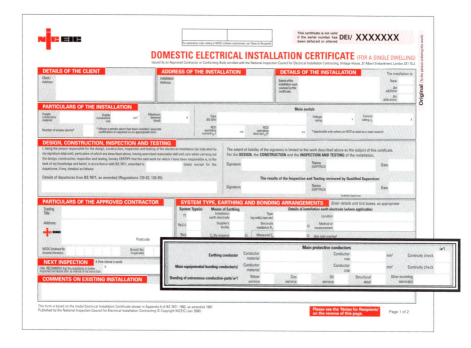

intended means of earthing is actually present and of the expected value, whatever the means of earthing provided. The external earth fault loop impedance is measured using an earth fault loop impedance test instrument (see Chapter 8).

Where the means of earthing is provided by an installation earth electrode, and where the earth electrode resistance has been measured using an earth fault loop impedance test instrument, the approximate ohmic value to be entered for Z_e will be the same as that recorded for R_A.

Main protective conductors

Earthing conductor

The earthing conductor is the protective conductor that connects the main earthing terminal of the installation (often located in the consumer unit) to the means of earthing. The conductor material and cross-sectional area of the earthing conductor must be stated here, followed by a tick to confirm that its continuity has been checked. This indication confirms that the inspector has tested the continuity of the earthing conductor, and/or inspected it throughout its length and checked its connections.

Main equipotential bonding conductors

The conductor material and cross-sectional area of the main bonding conductors must be stated here, followed by a tick to confirm that their continuity has been checked. This indication confirms that the inspector has tested the continuity of the bonding conductors, and/or inspected them throughout their length and checked their connections.

Bonding of extraneous-conductive-parts

Provision is made to record the extraneous-conductive-parts to which main bonding to the main earthing terminal of the installation has been provided, such as incoming water and gas services.

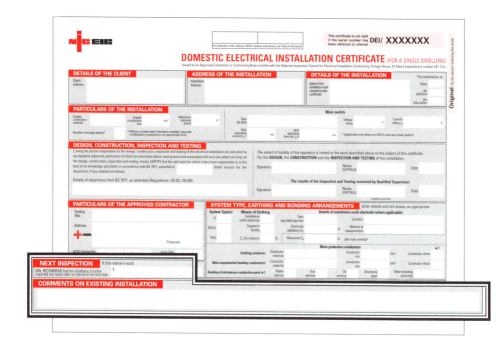

NEXT INSPECTION

This section should be completed by the person responsible for the installation work. A time interval, generally in terms of years, should be inserted to indicate when the next inspection will be due. *IEE Guidance Note 3: Inspection and Testing* gives guidance in terms of the maximum intervals between initial certification and the first periodic inspection and testing for various types of premises. For domestic purposes, the maximum interval to the first periodic inspection is normally ten years. The interval recorded on the certificate should take account of the available guidance material and any other special circumstances relating to the particular installation. Under no circumstances should 'N/A' (or 'Not Applicable') be inserted.

COMMENTS ON EXISTING INSTALLATION

This section of the certificate is applicable only where the work carried out is an addition or alteration to an existing installation, or where a complete rewire of an installation has been carried out and the installation is connected to an existing incoming supply. Any deficiencies observed in the existing installation which do not affect the safety of the new work must be recorded here. Examples of such deficiencies are deterioration and minor defects. Where there are no comments to be made, the entry should read 'None'.

It is important to appreciate that any defects which would result in a reduced level of safety in the new work (that is a level of safety less than would be afforded by compliance with *BS 7671*), must be corrected before the new work is put into service. Consequently, no such defects should be recorded on the Domestic Electrical Installation Certificate.

In particular, by the time the installation work is ready to be certified and put into service, the contractor should have ensured that the rating and condition of any existing equipment, such as the incoming supply, cables and consumer unit, which will have to carry any additional load as a result of the new work, are adequate for the altered circumstances and that the existing earthing arrangement, main equipotential bonding and, where appropriate, supplementary equipotential bonding are all adequate.

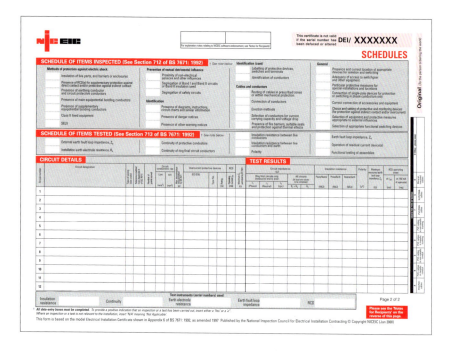

Chapter 3

SCHEDULES

The second page of the Domestic Electrical Installation Certificate comprises various schedules. Because use of the certificate is subject to particular limitations, it has been possible to omit those inspections and tests covered by the Electrical Installation Certificate which are not generally applicable to dwellings. It should be noted therefore that the Schedule of Items Inspected and the Schedule of Items Tested differ from those covered in Chapter 6. However, the guidance given in Chapter 6 on the completion of the schedules is largely applicable to the completion of the Electrical Installation Certificate, as is the guidance on the necessary inspections and tests given in Chapters 7 and 8, respectively.

Test instruments (serial numbers) used

Towards the foot of the page, a panel is provided for recording information about the test instruments used. The serial number of each of the instruments used to obtain the test results should be recorded here. If any instrument does not have a serial number, a suitable number should be assigned and permanently marked on the instrument for traceability purposes. Where a combined instrument such as an insulation/continuity test instrument is used to carry out more than one type of test, the serial number of that instrument should be listed in the space corresponding to each of the relevant types of test.

Inspection, Testing, Certification and Reporting

4

The Minor Electrical Installation Works Certificate

N|C EIC

This certificate is not valid if the serial number has been defaced or altered   **MEW/ XXXXXXX**

See reverse of this page for explanatory notes relating to NICEIC software endorsement

MINOR ELECTRICAL INSTALLATION WORKS CERTIFICATE (BS 7671: 1992)

Issued by an Approved Contractor or Conforming Body enrolled with the National Inspection Council for Electrical Installation Contracting, Vintage House, 37 Albert Embankment, London SE1 7UJ.

To be used only for minor electrical work which does not include the provision of a new circuit

PART 1: DETAILS OF THE MINOR WORKS

Details of departures, if any, from BS 7671: 1992 (as amended):

Client:

Date minor works completed:

Contract reference, if any:

Description of the minor works:

Location/address of the minor works:

PART 2: DETAILS OF THE MODIFIED CIRCUIT

| System type and earthing arrangements: | TN-C-S | | TN-S | TT | TN-C | IT | |

Method of protection against indirect contact:

| Overcurrent protective device for the modified circuit: | BS(EN) | | | Type | | Rating | | A |

| Residual current device (if applicable): | BS(EN) | | | Type | | $I_{\Delta n}$ | | mA |

Details of wiring system used to modify the circuit: Type   Reference method   csa of lives mm^2 csa of cpc mm^2

| Where protection against indirect contact is EEBAD: | Maximum disconnection time permitted by BS 7671 | s | Maximum Z_s permitted by BS 7671 | Ω |

Comments, if any, on existing installation:

PART 3: INSPECTION AND TESTING OF THE MODIFIED CIRCUIT AND RELATED PARTS   † Essential inspections and tests

† Confirmation that necessary inspections have been undertaken	(✓)	† Confirmation of the adequacy of earthing	(✓)
† Circuit resistance: $R_1 + R_2$ Ω or R_2 Ω		† Confirmation of the adequacy of equipotential bonding	(✓)
Insulation resistance: (* In a multi-phase circuit, record the lower or lowest value, as appropriate) Phase/Phase*	$M\Omega$	† Confirmation of correct polarity	(✓)
Phase/Neutral*	$M\Omega$	† Maximum measured earth fault loop impedance, Z_s	Ω
† Phase/Earth*	$M\Omega$	† RCD operating time at $I_{\Delta n}$ (if RCD fitted)	ms
† Neutral/Earth	$M\Omega$	RCD operating time at 150 mA, if applicable	ms

Agreed limitations, if any, on the inspection and testing:

PART 4: DECLARATION

I/We certify that the minor electrical installation works, as detailed in Part 1 of this certificate, does not impair the safety of the existing installation, that the said works have been designed, constructed, inspected, tested and verified in accordance with BS 7671:1992 (IEE Wiring Regulations), amended on the date shown* and that, to the best of my/our knowledge and belief, at the time of my/our inspection, the works complied with BS 7671:1992 except as detailed in Part 1 of this certificate. *

Name (CAPITALS)		For and on behalf of (Trading Title of Approved Contractor)	
Signature		Address and Postcode	
Position			
Date			

+ ■■■ Enrolment Number   Branch number (if applicable)   (The enrolment number is essential information)

This form is based on the model shown in Appendix 6 of BS 7671: 1992 as amended 1997
Published by the National Inspection Council for Electrical Installation Contracting © Copyright NICEIC (Jan 2000)

Please see the 'Notes for Recipients' on the reverse of this page.

Chapter 4

4 THE MINOR ELECTRICAL INSTALLATION WORKS CERTIFICATE

REQUIREMENTS

BS 7671 requires inspection and testing to be carried out, and appropriate certification to be issued, for any alteration or addition to an existing circuit, just as for a new electrical installation. However, provided that the alteration or addition to the installation is **minor** and does **not** include the provision of a new circuit, a Minor Electrical Installation Works Certificate (referred to in this Chapter as a Minor Works Certificate) may be issued instead of an Electrical Installation Certificate or a Domestic Electrical Installation Certificate.

A Minor Works Certificate may be used only for an addition or an alteration to a single circuit that does not extend to the provision of a new circuit.

The Minor Works Certificate therefore has a limited application and must not be used for work outside its scope. Examples of work that would fall within its scope are:

- The addition of a socket-outlet to a ring or radial final circuit.
- Work carried out on a lighting final circuit, including an alteration to the switching arrangements and the addition of a lighting point.
- The replacement of an accessory or luminaire.*

Further examples of proper use of Minor Works Certificates are given in the table on the following page.

Before commencing the minor electrical installation work, it is necessary to check that the rating and condition of any existing equipment, including that of the supplier (which may have to carry additional load), are adequate to accommodate in safety the altered circumstance resulting from the modifications to the circuit which is the subject of the certificate. Where the existence of a dangerous or potentially dangerous situation is discovered in the existing installation (such as the absence of main bonding where the method of protection against indirect contact is EEBAD), the alteration or addition should not proceed and the client should be advised immediately of the situation, preferably in writing, to satisfy the duties imposed on competent persons by the *Electricity at Work Regulations 1989.*

* If preferred by a client such as a Local Authority, the client's own documentation (such as a combined works order and certificate) may be used instead of a Minor Works Certificate for the replacement of an accessory or luminaire, provided that the safety declaration, installation details, and the inspection and test results required to be recorded on the alternative documentation are no less comprehensive than those required to be recorded on the Minor Works Certificate, and that a copy of the completed documentation is retained by the contractor.

ACCEPTABLE AND NON-ACCEPTABLE USE OF MINOR WORKS CERTIFICATES

Ref	Examples of electrical work	Acceptable use yes	Acceptable use no	Comment
A	Additional (non-emergency lighting) lighting points (luminaire and/or switching) on a single existing circuit.	✔		This acceptance is conditional on the existing circuit protective device being suitable to provide protection for the modified circuit, and other safety provisions being satisfactory.
B	Additional socket-outlets on a single existing radial or ring final circuit.	✔		This acceptance is conditional on the existing circuit protective device being suitable to provide protection for the modified circuit, and other safety provisions being satisfactory.
C	New circuit for lighting.		✘	This applies even where the circuit feeds only one lighting point.
D	New radial or ring final circuit for socket-outlets and other accessories such as connection units.		✘	This applies even where the circuit feeds only one socket-outlet or other accessory.
E	New radial circuit for fixed equipment.		✘	This applies even where the circuit feeds only one item of fixed equipment.
F	New circuit (lighting, power etc.) connected to an existing protective device.		✘	
G	Replacement of individual items of switchgear, including control switches and protective device(s), in a like-for-like manner for a single circuit.	✔		
H	Replacement of individual items of switchgear, including control switches and protective device(s), not in a like-for-like manner.		✘	
I	Replacement of main switchgear, incorporating protective devices, for more than one circuit.		✘	
J	Replacement of accessories, such as socket-outlets, ceiling roses on a like-for-like basis.	✔		Although essentially regarded as maintenance work, a record of the relevant safety tests undertaken may be recorded on the certificate though other suitable forms for recording such result are not precluded.
K	Installation of main equipotential bonding.	✔		
L	Upgrading main equipotential bonding.	✔		
M	Upgrading supplementary equipotential bonding.	✔		
N	Replacement of distribution cable, for a single circuit, damaged by, for example, fire, rodents impact etc.	✔		On condition that the replacement cable is identical in construction specification, and follows the same route as the damaged cable. Where more than one circuit is connected via a distribution board to the distribution circuit, the Minor Works Certificate is not suitable for test results.
O	Re-fixing, re-lidding existing wiring systems.	✔		On condition that the circuit's protective measures are unaffected.
P	Adding mechanical protection to existing equipment.	✔		On condition that the circuit's protective measures and current-carrying capacity of conductors, are unaffected.
Q	A combination of the above items, which may individually warrant the use of the certificate.		✘	The use of the Minor Works Certificate for certifying collectively more than one item, which individually may warrant the use of the certificate, is not acceptable.
R	Modifications to more than one circuit.		✘	
S	Periodic, or other, inspection of an installation.		✘	A single Minor Works Certificate should not be used for modifications to more than circuit.
T	For electrical installation work associated with fire detection and alarm systems.		✘	A Periodic Inspection Report form should be used
U	For electrical installation work associated with emergency lighting systems.		✘	
V	For electrical installation work carried out by others.		✘	
W	Portable appliance inspection and testing.		✘	

NOTE: This table does not represent an exhaustive list, but shows by example the limited scope of the Minor Works Certificate.

With regard to the replacement of accessories and luminaires, it is necessary to satisfy the requirements of *BS 7671*, no matter how minor the works may be. As a minimum, tests to confirm that shock protection has been provided are essential. The essential tests are earth fault loop impedance, polarity and, where an RCD is provided, the correct operation of the RCD. Where it is reasonably practicable, measurements of circuit impedance (R_1 + R_2) and/or R_2, and insulation resistance should also be carried out. The person signing the Minor Works Certificate is responsible for assessing each particular situation, and for deciding whether or not it is reasonable to omit any of the prescribed tests. In any event, a test should be omitted only with the express prior agreement of the client. Where one (or more) of the prescribed tests is not carried out, the contractor should give, on the Minor Works Certificate, brief details of the reasons and technical justifications for omitting the test.

USE AND MISUSE OF THE CERTIFICATE

Except where an Electrical Installation Certificate or a Domestic Electrical Installation Certificate is issued, a separate Minor Works Certificate (based on the model certificate prescribed by *BS 7671: 1992)* must be issued to the person who ordered the work, for **each** existing circuit on which minor works are carried out. This requirement applies whether or not the contractor is registered with the NICEIC, and whether or not certification has been specifically requested by the client.

As with the Electrical Installation Certificate and the Domestic Electrical Installation Certificate, the Minor Works Certificate is not intended to be issued to confirm the completion of a contract. The certificate is a declaration of electrical safety, and should be prepared before the alteration or addition to the circuit is put into service.

The NICEIC strongly prefers Approved Contractors to issue NICEIC certificates, as this provides a measure of confidence to the recipients. Non-approved contractors are not authorised to issue NICEIC certificates, but similar certificates based on the model given in *BS 7671* are available from the NICEIC.

An NICEIC Minor Works Certificate may be issued only by the Approved Contractor responsible for the installation work. The certificate must not be issued to cover an alteration or addition carried out by another contractor.

The table entitled 'Acceptable and Non-acceptable Use of Minor Works Certificates' shown opposite gives some examples of the circumstances in which the Minor Works Certificate may or may not be used. Most importantly, the certificate may be used only where the work is an alteration or an addition to an existing single circuit and where the work does not entail the provision of a new circuit, as clearly indicated towards the top of the certificate.

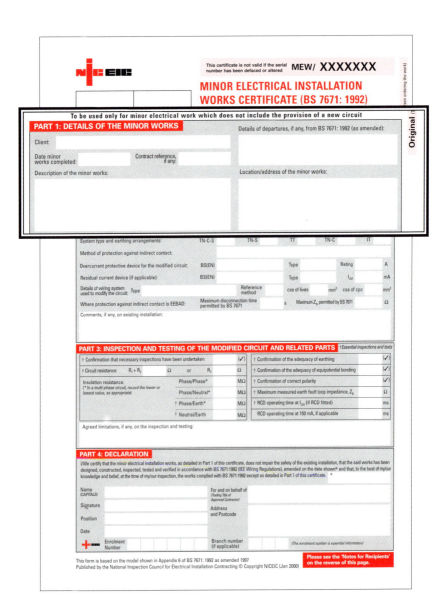

NICEIC

This certificate is not valid if the serial number has been defaced or altered **MEW/ XXXXXXX**

MINOR ELECTRICAL INSTALLATION WORKS CERTIFICATE (BS 7671: 1992)

Original

To be used only for minor electrical work which does not include the provision of a new circuit

PART 1: DETAILS OF THE MINOR WORKS

Details of departures, if any, from BS 7671: 1992 (as amended):

Client:

Date minor works completed:

Contract reference, if any:

Description of the minor works:

Location/address of the minor works:

System type and earthing arrangements: TN-C-S | TN-S | TT | TN-C | IT

Method of protection against indirect contact:

Overcurrent protective device for the modified circuit: BS(EN) | Type | Rating | A

Residual current device (if applicable): BS(EN) | Type | $I_{\Delta n}$ | mA

Details of wiring system used to modify the circuit: Type | Reference method | csa of lives | mm² csa of cpc | mm²

Where protection against indirect contact is EEBAD: Maximum disconnection time permitted by BS 7671 | s | Maximum Z_s permitted by BS 7671 | Ω

Comments, if any, on existing installation:

PART 3: INSPECTION AND TESTING OF THE MODIFIED CIRCUIT AND RELATED PARTS
† Essential inspections and tests

† Confirmation that necessary inspections have been undertaken	(✓)	† Confirmation of the adequacy of earthing	(✓)
† Circuit resistance: $R_1 + R_2$ Ω or R_n Ω		† Confirmation of the adequacy of equipotential bonding	(✓)
Insulation resistance: (* In a multi-phase circuit, record the lower or lowest value, as appropriate) Phase/Phase*	MΩ	† Confirmation of correct polarity	(✓)
Phase/Neutral*	MΩ	† Maximum measured earth fault loop impedance, Z_s	Ω
† Phase/Earth*	MΩ	† RCD operating time at $I_{\Delta n}$ (if RCD fitted)	ms
† Neutral/Earth	MΩ	RCD operating time at 150 mA, if applicable	ms

Agreed limitations, if any, on the inspection and testing:

PART 4: DECLARATION

I/We certify that the minor electrical installation works, as detailed in Part 1 of this certificate, does not impair the safety of the existing installation, that the said works has been designed, constructed, inspected, tested and verified in accordance with BS 7671:1992 (IEE Wiring Regulations), amended on the date shown* and that, to the best of my/our knowledge and belief, at the time of my/our inspection, the works complied with BS 7671:1992 except as detailed in Part 1 of this certificate. *

Name (CAPITALS)

For and on behalf of (Trading Title of Approved Contractor)

Signature

Address and Postcode

Position

Date

Enrolment Number

Branch number (if applicable) | (The enrolment number is essential information)

This form is based on the model shown in Appendix 6 of BS 7671: 1992 as amended 1997
Published by the National Inspection Council for Electrical Installation Contracting © Copyright NICEIC (Jan 2000)

Please see the 'Notes for Recipients' on the reverse of this page.

COMPILATION OF THE CERTIFICATE

The Minor Works Certificate has four parts. Guidance on the completion of each part is given in the following sections. Generally, irrespective of the method of compilation of the certificate, all unshaded data-entry boxes should be completed by inserting the necessary text, a positive indication such as a '✔' or a numeric value of the measured parameter, or by entering 'N/A' (meaning Not Applicable), where appropriate.

See 'Computer-assisted Preparation of Certificates and Reports' in Chapter 1 for guidance about the compilation of a Minor Works Certificate using a proprietary software package.

PART 1: DETAILS OF THE MINOR WORKS

This part of the certificate is where a clear description of the minor works is intended to be given. The minor electrical installation works should be clearly defined such that the work to which the certificate relates can be readily identified. All the data-entry boxes should be completed with the appropriate details that will precisely depict the work, its location and other relevant features, if any.

- **Client:** The name of the client should be stated in the space provided.

- **Date minor works completed:** The date the work was completed should be given.

- **Contract reference, if any:** The client's contract reference, if any, should be stated. In cases where no contract reference has been allocated, the data-entry box should record 'None'.

- **Description of the minor works:** A clear and accurate description of the minor works should be given here, as it is the safety aspects of this particular installation work for which the contractor is taking responsibility. **Failure to clearly and accurately identify the scope of the work could involve the contractor in unforeseen liabilities at a later date.**

- **Details of departures, if any, from BS 7671: 1992 (as amended):** Any design departures from *BS 7671* require special consideration by the designer (almost always the contractor in the case of work within the scope of this certificate). Design departures must not result in a degree of safety less than that provided by full compliance with *BS 7671*. Any such departures should be clearly identified in this part of the certificate and, in signing the declaration in Part 4, the contractor takes full responsibility for them. In cases where, as will generally be the case, no departures from *BS 7671* are sanctioned by the designer, the data-entry box should record 'None'.

NICEIC

This certificate is not valid if the serial number has been defaced or altered **MEW/ XXXXXXX**

MINOR ELECTRICAL INSTALLATION WORKS CERTIFICATE (BS 7671: 1992)

Issued by an Approved Contractor or Conforming Body enrolled with the National Inspection Council for Electrical Installation Contracting, Vintage House, 37 Albert Embankment, London SE1 7UJ.

See reverse of this page for explanatory notes relating to NICEIC software endorsement

Original (To the person ordering the work)

To be used only for minor electrical work which does not include the provision of a new circuit

PART 1: DETAILS OF THE MINOR WORKS

Client:

Details of departures, if any, from BS 7671: 1992 (as amended):

Date minor works completed:

Contract reference, if any:

Description of the minor works:

Location/address of the minor works:

PART 2: DETAILS OF THE MODIFIED CIRCUIT

System type and earthing arrangements: TN-C-S   TN-S   TT   TN-C   IT

Method of protection against indirect contact:

| Overcurrent protective device for the modified circuit: | BS(EN) | | Type | | Rating | A |

| Residual current device (if applicable): | BS(EN) | | Type | | $I_{\Delta n}$ | mA |

Details of wiring system used to modify the circuit: Type   Reference method   csa of lives mm^2 csa of cpc mm^2

Where protection against indirect contact is EEBAD: Maximum disconnection time permitted by BS 7671   s   Maximum Z_s permitted by BS 7671 Ω

Comments, if any, on existing installation:

† Confirmation that necessary inspections have been undertaken	(✓)	† Confirmation of the adequacy of earthing	(✓)
† Circuit resistance: $R_1 + R_2$ Ω or R_2 Ω		† Confirmation of the adequacy of equipotential bonding	(✓)
Insulation resistance: (* In a multi-phase circuit, record the lower or lowest value, as appropriate) Phase/Phase*	MΩ	† Confirmation of correct polarity	(✓)
† Phase/Neutral*	MΩ	† Maximum measured earth fault loop impedance, Z_s	Ω
† Phase/Earth*	MΩ	† RCD operating time at $I_{\Delta n}$ (if RCD fitted)	ms
† Neutral/Earth	MΩ	RCD operating time at 150 mA, if applicable	ms

Agreed limitations, if any, on the inspection and testing:

PART 4: DECLARATION

I/We certify that the minor electrical installation works, as detailed in Part 1 of this certificate, does not impair the safety of the existing installation, that the said works has been designed, constructed, inspected, tested and verified in accordance with BS 7671:1992 (IEE Wiring Regulations), amended on the date shown* and that, to the best of my/our knowledge and belief, at the time of my/our inspection, the works complied with BS 7671:1992 except as detailed in Part 1 of this certificate. *

Name (CAPITALS)		For and on behalf of (Trading Title of Approved Contractor)	
Signature		Address and Postcode	
Position			
Date			

Enrolment Number   Branch number (if applicable)   (The enrolment number is essential information)

This form is based on the model shown in Appendix 6 of BS 7671: 1992 as amended 1997
Published by the National Inspection Council for Electrical Installation Contracting © Copyright NICEIC (Jan 2000)

Please see the 'Notes for Recipients' on the reverse of this page.

- **Location/address of the minor works:** A clear and accurate description of the location and/or details of the address at which the minor works were executed should be recorded here.

PART 2: DETAILS OF THE MODIFIED CIRCUIT

Information concerning the modified circuit should be entered as follows:

- **System type and earthing arrangements:** The system type should be indicated by the insertion of a 'Yes' or a '✔' in one of the five boxes, as appropriate. Guidance as to the system type and earthing arrangements will be found in Chapter 2. In the United Kingdom, it is likely that the type of system will be TT, TN-S or TN-C-S. One data-entry box only must be completed to indicate the particular system type and earthing arrangement.

- **Method of protection against indirect contact:** This refers to the method used in the modified circuit to provide protection against indirect contact. The applicable method of protection against indirect contact for the modified circuit must have been selected from the five available options as set out in Regulation 413-01-01 of *BS 7671*. In many, if not most, cases this will be Earthed Equipotential Bonding and Automatic Disconnection of supply (EEBAD). The method used should be fully described here.

- **Overcurrent protective device for the modified circuit:** The information to be recorded here is the British Standard (or other appropriate Standard) product specification for the existing overcurrent protective device in terms of BS (EN) number, together with the type (if any) and the nominal current rating (I_n). For example, the type of device may be 'BS EN 60898 Type B circuit-breaker' and the nominal current rating in amperes may be '6' or '32' etc.

- **Residual current device (if applicable):** Where a residual current device already exists in the circuit or has been provided as part of the minor works for protection against electric shock, the information to be recorded here is the British Standard (or other appropriate Standard) product specification in terms of BS (EN) number, together with the type and the residual operating (tripping) current rating ($I_{\Delta n}$) in units of mA.

- **Details of wiring system used to modify the circuit:** The type of wiring system used should be identified here (eg: PVC cables in steel conduit) together with the reference method and cross-sectional areas of live and circuit protective conductors. Details of reference methods and associated installation methods are given in Table 4A of Appendix 4 of *BS 7671*.

NICEIC

This certificate is not valid if the serial number has been defaced or altered

MEW/ XXXXXXX

See reverse of this page for explanatory notes relating to NICEIC software endorsement

Original (To the person ordering the work)

MINOR ELECTRICAL INSTALLATION WORKS CERTIFICATE (BS 7671: 1992)

Issued by an Approved Contractor or Conforming Body enrolled with the National Inspection Council for Electrical Installation Contracting, Vintage House, 37 Albert Embankment, London SE1 7UJ.

To be used only for minor electrical work which does not include the provision of a new circuit

PART 1: DETAILS OF THE MINOR WORKS

Details of departures, if any, from BS 7671: 1992 (as amended):

Client:

Date minor works completed:

Contract reference, if any:

Description of the minor works:

Location/address of the minor works:

PART 2: DETAILS OF THE MODIFIED CIRCUIT

System type and earthing arrangements:	TN-C-S	TN-S	TT	TN-C	IT

Method of protection against indirect contact:

Overcurrent protective device for the modified circuit: BS(EN) — Type — Rating — A

Residual current device (if applicable): BS(EN) — Type — $I_{\Delta n}$ — mA

Details of wiring system used to modify the circuit: Type — Reference method — csa of lives — mm^2 — csa of cpc — mm^2

Where protection against indirect contact is EEBAD: Maximum disconnection time permitted by BS 7671 — s — Maximum Z_s permitted by BS 7671 — Ω

PART 3: INSPECTION AND TESTING OF THE MODIFIED CIRCUIT AND RELATED PARTS † *Essential inspections and tests*

† Confirmation that necessary inspections have been undertaken		(✔)
† Circuit resistance: $R_1 + R_2$ Ω or R_2		Ω
Insulation resistance: (* In a multi-phase circuit, record the lower or lowest value, as appropriate)	Phase/Phase*	$M\Omega$
	Phase/Neutral*	$M\Omega$
	† Phase/Earth*	$M\Omega$
	† Neutral/Earth	$M\Omega$

† Confirmation of the adequacy of earthing		(✔)
† Confirmation of the adequacy of equipotential bonding		(✔)
† Confirmation of correct polarity		(✔)
† Maximum measured earth fault loop impedance, Z_s		Ω
† RCD operating time at $I_{\Delta n}$ (if RCD fitted)		ms
RCD operating time at 150 mA, if applicable		ms

Agreed limitations, if any, on the inspection and testing:

PART 4: DECLARATION

I/We certify that the minor electrical installation works, as detailed in Part 1 of this certificate, does not impair the safety of the existing installation, that the said works has been designed, constructed, inspected, tested and verified in accordance with BS 7671:1992 (IEE Wiring Regulations), amended on the date shown* and that, to the best of my/our knowledge and belief, at the time of my/our inspection, the works complied with BS 7671:1992 except as detailed in Part 1 of this certificate. *

Name (CAPITALS)

For and on behalf of (Trading Title of Approved Contractor)

Signature

Address and Postcode

Position

Date

Enrolment Number

Branch number (if applicable)

(The enrolment number is essential information)

This form is based on the model shown in Appendix 6 of BS 7671: 1992 as amended 1997
Published by the National Inspection Council for Electrical Installation Contracting © Copyright NICEIC (Jan 2000)

Please see the 'Notes for Recipients' on the reverse of this page.

- **Maximum disconnection time permitted by BS 7671:** Where protection against indirect contact is by EEBAD (Earthed Equipotential Bonding and Automatic Disconnection of supply), the maximum disconnection time permitted by *BS 7671* should be recorded in the data-box provided, in units of seconds. This limit on disconnection time will be dependent on the type of circuit and its location.

- **Maximum Z_s permitted by BS 7671:** Where protection against indirect contact is by EEBAD, the maximum earth fault loop impedance, Z_s, permitted by *BS 7671* should be recorded in the data-entry box provided, in units of ohms.

- **Comments, if any, on existing installation:** The considerations relating to the completion of this element of the certificate are the same as those applicable to the section entitled 'comments on existing installation' in the Electrical Installation Certificate (see Chapter 2).

PART 3: INSPECTION AND TESTING OF THE MODIFIED CIRCUIT AND RELATED PARTS

The relevant provisions of Part 7 (Inspection and Testing) of *BS 7671* must be applied in full to electrical installation work covered by a Minor Works Certificate.

Generally, all data-entry boxes should be completed to confirm the results of a particular inspection or test by a 'Yes' or a '✔', or by the insertion of a measured value where appropriate. Where a particular inspection or test is not applicable, this should be indicated by inserting 'N/A', meaning Not Applicable. Exceptionally, where an inspection or a test is not practicable, the entry should be 'LIM' meaning 'Limitation', indicating that the particular circumstances prevented such an inspection or test procedure from being carried out. In such cases, the limitation(s) should be agreed with the client before the work is undertaken, and identified in the space provided at the bottom of Part 3. The inspections and tests identified on the certificate with the mark '†' are considered essential for confirming the safety of all minor electrical installation work involving the alteration or addition of cables and conductors.

The method for carrying out each item of inspection and testing is described in Chapter 7 and Chapter 8, respectively.

The information required for Part 3 of the Minor Works Certificate is as follows:

- **Confirmation that necessary inspections have been undertaken:** A positive entry such as a tick (✔) is necessary here to indicate that all necessary inspections have been carried out with satisfactory outcomes, as appropriate to the particular circuit modification.

- **Circuit resistance $R_1 + R_2$, or R_2:** Enter the highest measured value of $R_1 + R_2$, in ohms, measured at the each of the points and/or accessories on the modified circuit. During this test, the phase and circuit protective conductors of the circuit

N|C|EIC

See reverse of this page for explanatory notes relating to NICEIC software endorsement

This certificate is not valid if the serial number has been defaced or altered **MEW/ XXXXXXX**

MINOR ELECTRICAL INSTALLATION WORKS CERTIFICATE (BS 7671: 1992)

Issued by an Approved Contractor or Conforming Body enrolled with the National Inspection Council for Electrical Installation Contracting, Vintage House, 37 Albert Embankment, London SE1 7UJ.

Original (To the person ordering the work)

To be used only for minor electrical work which does not include the provision of a new circuit

PART 1: DETAILS OF THE MINOR WORKS

Client:

Date minor works completed:

Contract reference, if any:

Description of the minor works:

Details of departures, if any, from BS 7671: 1992 (as amended):

Location/address of the minor works:

PART 2: DETAILS OF THE MODIFIED CIRCUIT

System type and earthing arrangements:	TN-C-S	TN-S	TT	TN-C	IT

Method of protection against indirect contact:

Overcurrent protective device for the modified circuit:	BS(EN)		Type	Rating	A
Residual current device (if applicable):	BS(EN)		Type	$I_{\Delta n}$	mA

Details of wiring system used to modify the circuit:	Type		Reference method	csa of lives mm^2	csa of cpc mm^2

| Where protection against indirect contact is EEBAD: | Maximum disconnection time permitted by BS 7671 | s | Maximum Z_s permitted by BS 7671 | Ω |

Comments, if any, on existing installation:

PART 3: INSPECTION AND TESTING OF THE MODIFIED CIRCUIT AND RELATED PARTS   † Essential inspections and tests

† Confirmation that necessary inspections have been undertaken	(✓)	† Confirmation of the adequacy of earthing	(✓)
† Circuit resistance: $R_1 + R_2$ Ω or R_2	Ω	† Confirmation of the adequacy of equipotential bonding	(✓)
Insulation resistance: (* In a multi-phase circuit, record the lower or lowest value, as appropriate) Phase/Phase*	$M\Omega$	† Confirmation of correct polarity	(✓)
Phase/Neutral*	$M\Omega$	† Maximum measured earth fault loop impedance, Z_s	Ω
† Phase/Earth*	$M\Omega$	† RCD operating time at $I_{\Delta n}$ (if RCD fitted)	ms
† Neutral/Earth	$M\Omega$	RCD operating time at 150 mA, if applicable	ms

Agreed limitations, if any, on the inspection and testing:

PART 4: DECLARATION

I/We certify that the minor electrical installation works, as detailed in Part 1 of this certificate, does not impair the safety of the existing installation, that the said works has been designed, constructed, inspected, tested and verified in accordance with BS 7671:1992 (IEE Wiring Regulations), amended on the date shown* and that, to the best of my/our knowledge and belief, at the time of my/our inspection, the works complied with BS 7671:1992 except as detailed in Part 1 of this certificate. *

Name (CAPITALS)		For and on behalf of (Trading Title of Approved Contractor)	
Signature		Address and Postcode	
Position			
Date			

╬=▪▪▪ Enrolment Number

Branch number (if applicable)

(The enrolment number is essential information)

This form is based on the model shown in Appendix 6 of BS 7671: 1992 as amended 1997
Published by the National Inspection Council for Electrical Installation Contracting © Copyright NICEIC (Jan 2000)

Please see the 'Notes for Recipients' on the reverse of this page.

should be connected together as described for measurement of $R_1 + R_2$ in Chapter 8 (see 'continuity of protective conductors' or, for ring final circuits, 'continuity of ring final circuit conductors'). This test procedure also confirms that polarity is correct and gives an ohmic value of the circuit's contribution to the earth fault loop impedance Z_s. Alternatively, if preferred, the impedance of the circuit protective conductor R_2 may be recorded instead of $R_1 + R_2$, but in this case correct polarity will need to be separately confirmed by other means (see Chapter 8).

- **Insulation resistance:** Enter the measured values of phase/phase (only applicable for multi-phase circuits), phase/neutral, phase/earth and neutral/earth insulation resistance of the circuit, in megohms. Where the circuit is multi-phase (eg two-phase or three-phase), the lower or lowest value of measured insulation resistance should be recorded. See Chapter 8 regarding precautions that may be necessary before testing insulation resistance.

- **Confirmation of the adequacy of earthing:** A positive entry such as a tick (✔) is necessary here to indicate that the adequacy of the earthing has been established by inspection of the earthing conductor and its terminations, and by measurement of the earth fault loop impedance (Z_e), both at the origin of the installation.

- **Confirmation of the adequacy of equipotential bonding:** A positive entry such as a tick (✔) is necessary here to confirm that the adequacy of the equipotential bonding to all extraneous-conductive-parts has been established by inspection and by continuity checks of the bonding conductors and their terminations.

- **Confirmation of correct polarity:** Correct polarity will have been confirmed during the $R_1 + R_2$ test procedure or, if the R_2 measurement was taken instead, by other means (described in Chapter 8). A positive indication such as a tick (✔) entered in this part of the certificate signifies that correct polarity has been confirmed.

- **Maximum measured earth fault loop impedance, Z_s:** Enter the highest value of earth fault loop impedance, Z_s, in ohms, measured at each of the points and/or accessories on the modified circuit. The highest measured value of Z_s must be such as to afford automatic disconnection within the time limit prescribed in *BS 7671*.

- **RCD operating time at $I_{\Delta n}$ (if RCD fitted):** If there is an RCD in circuit, this should be tested in the manner described in Chapter 8. The operating (tripping) time in milliseconds (ms) when subjected to a test at current equal to $I_{\Delta n}$ should be recorded in the space provided.

N|C EIC

This certificate is not valid if the serial number has been defaced or altered **MEW/ XXXXXXX**

See reverse of this page for explanatory notes relating to NICEIC software endorsement

MINOR ELECTRICAL INSTALLATION WORKS CERTIFICATE (BS 7671: 1992)

Issued by an Approved Contractor or Conforming Body enrolled with the National Inspection Council for Electrical Installation Contracting, Vintage House, 37 Albert Embankment, London SE1 7UJ.

Original (To the person ordering the work)

To be used only for minor electrical work which does not include the provision of a new circuit

PART 1: DETAILS OF THE MINOR WORKS

Client:

Date minor works completed:

Contract reference, if any:

Description of the minor works:

Details of departures, if any, from BS 7671: 1992 (as amended):

Location/address of the minor works:

PART 2: DETAILS OF THE MODIFIED CIRCUIT

System type and earthing arrangements: TN-C-S TN-S TT TN-C IT

Method of protection against indirect contact:

Overcurrent protective device for the modified circuit: BS(EN) ___ Type ___ Rating ___ A

Residual current device (if applicable): BS(EN) ___ Type ___ $I_{\Delta n}$ ___ mA

Details of wiring system used to modify the circuit: Type ___ Reference method ___ csa of lives ___ mm² csa of cpc ___ mm²

Where protection against indirect contact is EEBAD: Maximum disconnection time permitted by BS 7671 ___ s Maximum Z_s permitted by BS 7671 ___ Ω

Comments, if any, on existing installation:

PART 3: INSPECTION AND TESTING OF THE MODIFIED CIRCUIT AND RELATED PARTS †Essential inspections and tests

† Confirmation that necessary inspections have been undertaken	(✓)	
† Circuit resistance: $R_1 + R_2$ ___ Ω or R_2 ___ Ω		
Insulation resistance: (* In a multi-phase circuit, record the lower or lowest value, as appropriate)	Phase/Phase* ___ MΩ	
	Phase/Neutral* ___ MΩ	
† Phase/Earth* ___ MΩ		
† Neutral/Earth ___ MΩ		

† Confirmation of the adequacy of earthing (✓)
† Confirmation of the adequacy of equipotential bonding (✓)
† Confirmation of correct polarity (✓)
† Maximum measured earth fault loop impedance, Z_s ___ Ω
† RCD operating time at $I_{\Delta n}$ (if RCD fitted) ___ ms
RCD operating time at 150 mA, if applicable ___ ms

Agreed limitations, if any, on the inspection and testing:

PART 4: DECLARATION

I/We certify that the minor electrical installation works, as detailed in Part 1 of this certificate, does not impair the safety of the existing installation, that the said works has been designed, constructed, inspected, tested and verified in accordance with BS 7671:1992 (IEE Wiring Regulations), amended on the date shown* and that, to the best of my/our knowledge and belief, at the time of my/our inspection, the works complied with BS 7671:1992 except as detailed in Part 1 of this certificate. *

Name (CAPITALS):

Signature:

Position:

Date:

For and on behalf of (Trading Title of Approved Contractor):

Address and Postcode:

Enrolment Number:

Branch number (if applicable):

(The enrolment number is essential information)

This form is based on the model shown in Appendix 6 of BS 7671: 1992 as amended 1997
Published by the National Inspection Council for Electrical Installation Contracting © Copyright NICEIC (Jan 2000)

Please see the 'Notes for Recipients' on the reverse of this page.

- **RCD operating time at 150 mA, (if applicable):** If there is an RCD in circuit which is intended to provide supplementary protection against direct contact, it should be tested in the manner described in Chapter 8. The operating (tripping) time in milliseconds (ms) when subjected to a test at 150 mA should be recorded in the space provided. If the RCD is not intended to provide supplementary protection against direct contact the test is not essential, and if not undertaken, insert 'N/A' (meaning Not Applicable) in the box.

- **Agreed limitations, if any, on the inspection and testing:** All limitations imposed on the inspection and testing should be explained to, and agreed with, the client before work commences. Such limitations should be recorded in the space provided together with the technical justifications for taking such decisions. **It should be appreciated that failure to clearly explain and record all of the agreed limitations of the inspection and testing could involve the contractor in unforeseen liabilities at a later date.**

With regard to the replacement of accessories and luminaires, it may be impracticable to carry out all the identified tests. However, testing of earth fault loop impedance, polarity and, where an RCD is provided, confirmation of the correct operation of the RCD are considered to be essential. Where practicable, circuit impedance ($R_1 + R_2$) and/or R_2, insulation resistance and earth fault loop impedance tests should also be carried out.

For example, where a socket-outlet is added to an existing circuit, it is necessary to:

- Carry out all necessary inspecting and testing of the modified circuit and related parts of the installation and supply on which the circuit depends for protection. This includes inspecting and testing to confirm that the earthing and equipotential bonding arrangements for the modified circuit are both adequate and reliable.

- Check the continuity and resistance of protective conductor(s) to establish that the earthing contact of the socket-outlet is reliably and effectively connected to the main earthing terminal of the installation via a low impedance circuit protective conductor.

- Measure the insulation resistances, phase to phase, phase to neutral, and phase to earth of the circuit that has been modified, and establish that the resistance values comply with Table 71A of *BS 7671*.

- Check that the polarity at the socket-outlet is correct.

- Measure the maximum earth fault loop impedance to establish that the permitted disconnection time is not exceeded - see table provided with NICEIC Minor Works Certificates.

- Check the suitability and effectiveness of the RCD, if the modified circuit is so protected.

NICEIC

This certificate is not valid if the serial number has been defaced or altered **MEW/ XXXXXXX**

See reverse of this page for explanatory notes relating to NICEIC software endorsement

MINOR ELECTRICAL INSTALLATION WORKS CERTIFICATE (BS 7671: 1992)

Issued by an Approved Contractor or Conforming Body enrolled with the National Inspection Council for Electrical Installation Contracting, Vintage House, 37 Albert Embankment, London SE1 7UJ.

Original (To the person ordering the works)

To be used only for minor electrical work which does not include the provision of a new circuit

PART 1: DETAILS OF THE MINOR WORKS

Client:

Date minor works completed:

Contract reference, if any:

Description of the minor works:

Details of departures, if any, from BS 7671: 1992 (as amended):

Location/address of the minor works:

PART 2: DETAILS OF THE MODIFIED CIRCUIT

System type and earthing arrangements:	TN-C-S	TN-S	TT	TN-C	IT

Method of protection against indirect contact:

Overcurrent protective device for the modified circuit: BS(EN)   Type   Rating   A

Residual current device (if applicable): BS(EN)   Type $I_{\Delta n}$ mA

Details of wiring system used to modify the circuit: Type   Reference method   csa of lives   mm²   csa of cpc   mm²

Where protection against indirect contact is EEBAD: Maximum disconnection time permitted by BS 7671   s   Maximum Z_s permitted by BS 7671   Ω

Comments, if any, on existing installation:

PART 3: INSPECTION AND TESTING OF THE MODIFIED CIRCUIT AND RELATED PARTS   † Essential inspections and tests

† Confirmation that necessary inspections have been undertaken	(✓)	† Confirmation of the adequacy of earthing	(✓)
† Circuit resistance: $R_1 + R_2$ Ω or R_2 Ω		† Confirmation of the adequacy of equipotential bonding	(✓)
Insulation resistance: (* In a multi-phase circuit, record the lower or lowest value, as appropriate)	Phase/Phase* MΩ	† Confirmation of correct polarity	(✓)
	Phase/Neutral* MΩ	† Maximum measured earth fault loop impedance, Z_s	Ω
	† Phase/Earth* MΩ	† RCD operating time at $I_{\Delta n}$ (if RCD fitted)	ms
	† Neutral/Earth MΩ	RCD operating time at 150 mA, if applicable	ms

PART 4: DECLARATION

I/We certify that the minor electrical installation works, as detailed in Part 1 of this certificate, does not impair the safety of the existing installation, that the said works has been designed, constructed, inspected, tested and verified in accordance with BS 7671:1992 (IEE Wiring Regulations), amended on the date shown* and that, to the best of my/our knowledge and belief, at the time of my/our inspection, the works complied with BS 7671:1992 except as detailed in Part 1 of this certificate.   *

Name (CAPITALS)

Signature

Position

Date

For and on behalf of (Trading Title of Approved Contractor)

Address and Postcode

Enrolment Number

Branch number (if applicable)   (The enrolment number is essential information)

This form is based on the model shown in Appendix 6 of BS 7671: 1992 as amended 1997
Published by the National Inspection Council for Electrical Installation Contracting © Copyright NICEIC (Jan 2000)

Please see the 'Notes for Recipients' on the reverse of this page.

PART 4: DECLARATION

In this part of the certificate, the contractor certifies that the work undertaken in connection with the alteration or addition does not impair the safety of the existing installation, and that the new work has been inspected and tested in accordance with *BS 7671*. The contractor also certifies that the alteration or addition complies with the requirements of *BS 7671* (as amended), with the exception of any design departures identified in Part 1 of the certificate.

BS 7671 Amendment: The date of the amendment to *BS 7671* that was current at the date when the work was carried out should be stated in the relevant space in the text of the declaration.

Name (CAPITALS): The name of the competent person providing the declaration on behalf of the contractor should be recorded in capital letters in the space provided.

Signature: The signature of the competent person providing the declaration on behalf of the contractor, and the date of signing, should be inserted in the appropriate spaces. The signatory should be a competent person who is well informed about the design and the construction of the work. The signatory should have personally carried out and verified the work, or have undertaken detailed supervision of the work, so as to be in a position to take with confidence full responsibility for the statements made in the declaration.

Position: The position of the signatory should be identified in the space provided (eg: NICEIC Qualified Supervisor, director, partner, or supervising electrical engineer).

Date: The date on which the signatory signs the declaration should be recorded in the space provided.

For and on behalf of (Trading Title of Approved Contractor): The Approved Contractor's full trading title should be inserted in the space provided.

Address and Postcode: The Approved Contractor's full postal address, including the postcode, should be inserted in the space provided.

NICEIC Enrolment Number: It is essential that this information is recorded in the space provided, together with NICEIC Branch number where appropriate.

Inspection, Testing, Certification and Reporting

Chapter 5

The Periodic Inspection Report

NICEIC

This report is not valid if the serial number has been defaced or altered

PIR/ XXXXXXX

Original (To the person ordering the work)

See reverse of this page for explanatory notes relating to NICEIC software endorsement

PERIODIC INSPECTION REPORT
FOR AN ELECTRICAL INSTALLATION

Issued by an Approved Contractor or Conforming Body enrolled with the National Inspection Council for Electrical Installation Contracting, Vintage House, 37 Albert Embankment, London SE1 7UJ.

A. DETAILS OF THE CLIENT

Client:

Address:

B. PURPOSE OF THE REPORT

This Periodic Inspection Report must be used only for reporting on the condition of an existing installation.

Purpose for which this report is required:

C. DETAILS OF THE INSTALLATION

Domestic   Commercial   Industrial

Occupier:

Description of premises:

Address:

Other: (Please state)

Estimated age of the electrical installation: years

Postcode:

Evidence of alterations or additions

If yes, estimated age   years

Date of previous inspection:

Electrical Installation Certificate No or previous Periodic Inspection Report No:

Records of installation available:

Records held by:

D. EXTENT OF THE INSTALLATION AND LIMITATIONS OF THE INSPECTION AND TESTING

Extent of the electrical installation covered by this report:

Agreed limitations, if any, on the inspection and testing:

This inspection has been carried out in accordance with BS 7671: 1992 (IEE Wiring Regulations), as amended. Cables concealed within trunking and conduits, or cables and conduits concealed under floors, in inaccessible roof spaces and generally within the fabric of the building or underground, have not been visually inspected.

E. DECLARATION

I/We, being the person(s) responsible for the inspection and testing of the electrical installation (as indicated by my/our signatures below), particulars of which are described above (see C), having exercised reasonable skill and care when carrying out the inspection and testing, hereby declare that the information in this report, including the observations (see F) and the attached schedules (see H), provides an accurate assessment of the condition of the electrical installation taking into account the stated extent of the installation and the limitations of the inspection and testing (see D). I/We further declare that in my/our judgement, the said installation was overall in ✦ _____ condition (see G) at the time the inspection was carried out, and that it should be further inspected as recommended (see I).

✦ (Insert 'a satisfactory' or 'an unsatisfactory', as appropriate)

INSPECTION, TESTING AND ASSESSMENT BY:

Signature:

Name: (CAPITALS)

Position:

Date:

REPORT REVIEWED AND CONFIRMED BY: † See note below

Signature:

Name: (CAPITALS)

(Registered Qualified Supervisor for the Approved Contractor at J)

Date:

Page 1 of

† This Periodic Inspection Report should be reviewed and confirmed by the registered Qualified Supervisor for the Approved Contractor responsible for issuing the Report.

This form is based on the model shown in Appendix 6 of BS 7671: 1992 as amended 1997
Published by the National Inspection Council for Electrical Installation Contracting © Copyright NICEIC (Jan 2000)

Please see the 'Notes for Recipients' on the reverse of this page.

⑤ THE PERIODIC INSPECTION REPORT

THE NEED FOR PERIODIC INSPECTION AND TESTING

Every electrical installation deteriorates with usage and age. It is important for the person responsible for the maintenance of the installation to be sure that the safety of the users is not put at risk, and that the installation continues to be in a safe and serviceable condition. It is necessary therefore for the installation to be periodically inspected and tested and a report on its condition obtained. Deficiencies observed during the inspection and testing may then be remedied such that the installation may continue to be used in safety.

The periodic inspection of an electrical installation may be required for one or more of a variety of reasons, each of which may impose particular requirements or limitations.

PURPOSE OF PERIODIC INSPECTION

The main purpose of periodic inspection and testing is to detect so far as is reasonably practicable, and to report on, any factors impairing the safety of an electrical installation. The aspects to be covered, as stated in Regulation 731-01-02, include all of the following:

- Safety of persons and livestock against the effects of electric shock and burns.
- Protection against damage to property by fire and heat from an installation defect.
- Confirmation that the installation is not damaged or deteriorated so as to impair safety.
- Identification of non-compliances with *BS 7671* or installation defects which might give rise to danger.

Unless the circumstances make it unavoidable (for example, if an installer has ceased trading prior to certifying an installation), a Periodic Inspection Report should not be issued by one contractor as a substitute for an Electrical Installation Certificate or a Domestic Electrical Installation Certificate for work carried out by another contractor. A Periodic Inspection Report does not provide a declaration by the designer or installer that the aspects of the work for which they were responsible comply with *BS 7671*. Also, cables that are designed to be concealed cannot be inspected when construction is complete.

If a contractor is called upon to issue a Periodic Inspection Report for work designed and installed by others, it would be prudent for that contractor to make the purpose and limitations of the report absolutely clear, in order to avoid unwittingly assuming responsibility for aspects of the work of which the contractor had no knowledge and over which the contractor had no control.

Recommended Initial Frequencies of Inspection of Electrical Installations

Type of installation	Maximum period between inspections and testing as necessary	Reference (see notes below)
General installation		
Domestic	Change of tenancy/10 years	
Commercial	Change of tenancy/5 years	1, 2
Educational establishments	5 years	1, 2
Hospitals	5 years	1, 2
Industrial	3 years	1, 2
Residential accommodation	5 years	1
Offices	5 years	1, 2
Shops	5 years	1, 2
Laboratories	5 years	1, 2
Buildings open to the public		
Cinemas	1 year	2, 6, 7
Church installations	5 years	2
Leisure complexes	1 year	1, 2, 6
Places of public entertainment	1 year	1, 2, 6
Restaurants and hotels	5 years	1, 2, 6
Theatres	1 year	2, 6, 7
Public houses	5 years	1, 2, 6
Village/Community halls	5 years	1, 2
External installations		
Agricultural & horticultural	3 years	1, 2
Caravans	3 years	
Caravan parks	1 year	1, 2, 6
Highway power supplies	6 years	
Marinas	1 years	1, 2
Fish farms	1 years	1, 2
Special installations		
Emergency lighting	3 years	2, 3, 4
Fire alarms	1 year	2, 4, 5
Launderettes	1 year	1, 2, 6
Petrol filling stations	1 year	1, 2, 6
Construction site installations	3 months	1, 2

Reference Key

1. Particular attention must be taken to comply with SI 1988 No 1057. The Electricity Supply Regulations 1988 (as amended).
2. SI 1989 No 635. The Electricity at Work Regulations 1989 (Regulation 4 & Memorandum).
3. See BS 5266: Part 1: 1999 Code of practice for the emergency lighting of premises other than cinemas and certain other specified premises used for entertainment.
4. Other intervals are recommended for testing operation of batteries and generators.
5. See BS 5839: Part 1: 1988 Code of practice for system design installation and servicing (Fire detection and alarm systems for buildings).
6. Local Authority Conditions of Licence.
7. SI 1995 No 1129 (Clause 27) The Cinematography (Safety) Regulations.

Table reproduced by kind permission of the Institution of Electrical Engineers

SAFETY

Attention is drawn to the need for all persons involved in inspection and testing to be competent, and to follow safe working procedure - see Chapter 1

If the absence of sufficient installation information prevents the inspection and testing being carried out safely, information (such as diagrams and circuit details) must be obtained from the person responsible for the installation. Alternatively, the information must be prepared for the purpose before work commences, in order to afford compliance with Section 6 of the *Health and Safety at Work etc Act 1974*.

INTERVALS BETWEEN PERIODIC INSPECTION AND TESTING

The interval between periodic inspection and testing will depend on a number of considerations, including some or all of the factors indicated below, depending on the particular circumstances:

- Age of the installation (it may be reasonable to expect the intervals to become progressively shorter as the installation ages).
- Type of premises.
- Environmental conditions (ie external influences).
- Normal life expectancy of the installation.
- Level of misuse of the installation (eg vandalism).
- Changed usage of the premises (and the installation).
- The extent of any wear and tear, damage or other deterioration.

Determination of the interval to the next inspection will always be a matter of engineering judgement to be exercised by the inspector. However, the table opposite (by kind permission of the Institution of Electrical Engineers) gives recommendations for the maximum initial interval between initial verification and the first periodic inspection. The information given should be used as a guide only and the intervals stated may need to be adjusted to meet the particular circumstances.

REQUIREMENTS

BS 7671 requires that a Periodic Inspection Report (which includes a schedule of test results) is issued following the in-service periodic inspection and testing of an electrical installation. A separate report is required for each distinct installation examined. The report provides a formal declaration that, within stated and agreed limitations, the details recorded, including the observations and recommendations and the completed schedules of inspection and test results, give an accurate assessment of the condition of the electrical installation.

Inspection, Testing, Certification and Reporting

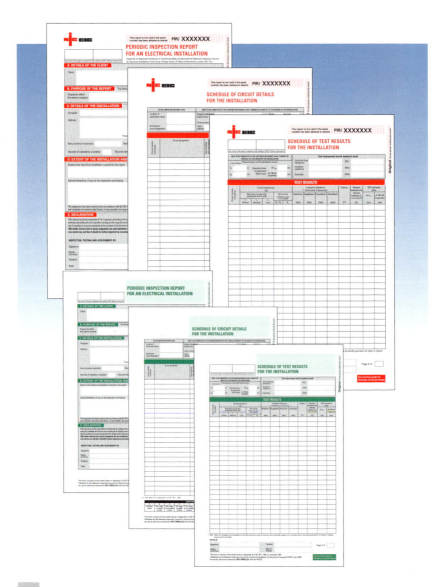

Wherever possible, before undertaking periodic inspection and testing, it is important for the person undertaking the work of inspection to have sight of the records associated with the installation. Such records would include the Electrical Installation Certificate (formerly the Electrical Installation Completion Certificate) or a Domestic Electrical Installation Certificate issued when the installation was originally completed and any subsequent Minor Electrical Installation Works Certificates and Periodic Inspection Reports. Depending on the complexity of the installation, the records might also include 'as-fitted' drawings and single-line distribution drawings.

A report (based on the standard form prescribed by *BS 7671*) should be issued to the person ordering the inspection and test, whether or not the contractor is enrolled with the NICEIC, and whether or not a written report has been specifically requested by the client. The NICEIC strongly prefers Approved Contractors to issue NICEIC report forms, as this provides a measure of confidence to the recipients. Non-approved contractors are not authorised to issue NICEIC Periodic Inspection Report forms, but similar report forms based on the model form given in *BS 7671* are available from the NICEIC.

The Periodic Inspection Report is, as its title indicates, a report and not a certificate. It relates to an assessment of the in-service condition of an electrical installation against the requirements of the issue of *BS 7671* current at the time of the inspection, irrespective of the age of the installation. The criteria for assessing the compliance of each part of an older installation with *BS 7671* are, therefore, the same as for new installation work. The report is for the benefit of the person ordering the work, and of persons subsequently involved in additional or remedial work, or further inspections.

The Periodic Inspection Report form is to be used only for reporting on the condition of an existing installation. It must not be used instead of an Electrical Installation Certificate or a Domestic Electrical Installation Certificate for certifying a new electrical installation work, or as a substitute for a Minor Electrical Installation Works Certificate for certifying an addition or an alteration to an existing installation.

For installations having more than one distribution board, or more circuits than can be recorded on Pages 5 and 6, one or more additional pages of the Schedule of Circuit Details for the Installation and the Schedule of Test Results for the Installation will be required. Continuation schedules are available separately from the NICEIC. The additional pages are to be given the same unique serial number as the other pages of the report, by first striking out 'EIC' and then adding the remainder of the unique serial number in the space allocated (see the following page). Only the version of the continuation schedules intended for use in conjunction with the current version of the Periodic Inspection Report should be used. Superseded versions of continuation schedules must not be used. The page number for each additional schedule is to be inserted, together with the total number of pages comprising the report (eg Page 7 of 8).

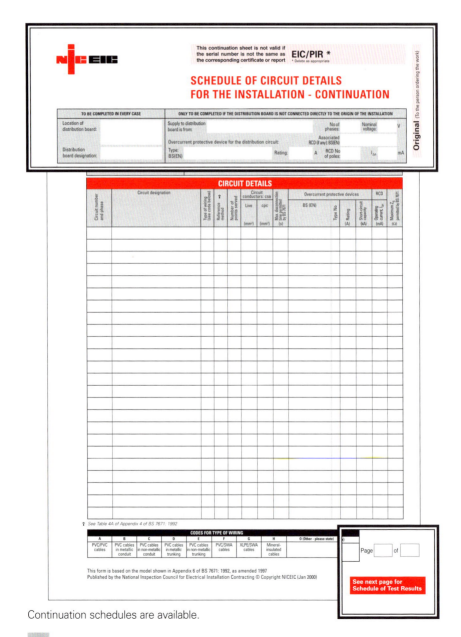

Continuation schedules are available.

Generally, irrespective of whether the method of compilation of the report is by hand or aided by computer software, all unshaded data-entry boxes must be completed by inserting the necessary text, or a 'Yes' or a '✔' to indicate that the task has been completed, or a numeric value of the measured parameter, or by entering 'N/A' meaning 'Not Applicable', where appropriate. Exceptionally, where a limitation on a particular inspection or test has been agreed, and has been recorded in Section D, the appropriate data-entry box(es) should be completed by inserting 'LIM', indicating that an agreed limitation has prevented the inspection or test being carried out.

The total number of pages which make up the report must be inserted in the box provided at the foot of each of the pages on the right-hand side.

PERIODIC INSPECTION AND TESTING PROCEDURES

The procedures for periodic inspection and testing differ in some respects from those for the initial verification of new installation work. This is because the subject of a periodic inspection report is usually an installation which has been energised and in use for some time, and particular attention therefore needs to be given during the inspection process to assessing the condition of the installation in respect of:

- Safety.
- Corrosion.
- Excessive loading.
- External influences.
- Wear and tear.
- Damage and deterioration.
- Age.
- Suitability (taking account of any changes in use or building layout etc).

Also, for reasons beyond the inspector's control, the inspector may be unable to gain access to parts of the existing installation. For example, it is usually impossible to inspect cables that have been concealed within the fabric of the building. It may also not be permitted to switch off certain parts of an existing installation due to operational restrictions imposed by the client. Such restrictions are likely to result in the inspection and testing of those parts of the installation being very limited, or the parts may have to be omitted entirely from the process.

Some degree of sampling may be involved in the inspection and testing process. Such sampling requires careful consideration being given to the selection of the parts of the installation to be inspected and tested. In the event of deficiencies being discovered in the initial sample, the size of the sample should be increased as described in *IEE Guidance Note 3*. The sampling technique used, and the parts of the installation sampled, should be carefully recorded so that a different sample can be selected for the next and subsequent periodic inspections.

Sequence of Tests	
For initial testing	**For periodic testing**
(i) Before the supply is connected, or with the supply disconnected as appropriate	(i) The following are generally applicable
• Continuity of protective conductors, main and supplementary bonding.	• Continuity of all protective conductors (including earthed equipotential bonding conductors).
• Continuity of ring final circuit conductors.	• Polarity.
• Insulation resistance.	• Earth fault loop impedance.
• Site-applied insulation.	• Insulation resistance.
• Protection by separation of circuits.	• Operation of devices for isolation and switching.
• Protection by barriers or enclosures provided during erection.	• Operation of residual current devices.
• Insulation of non-conducting floors and walls.	• Prospective fault current.
• Polarity.	
• Earth electrode resistance.	(ii) Where appropriate, the following tests must also be undertaken
	• Continuity of ring circuit conductors.
(ii) With the electrical supply connected (re-check polarity before further testing)	• Earth electrode resistance.
• Prospective fault current.	• Manual operation of overcurrent protective devices other than fuses.
• Earth fault loop impedance.	• Electrical separation of circuits - insulation resistance.
• Residual current operated devices.	• Protection by non-conducting floors and walls - insulation resistance.
• Functional test of switchgear and controlgear.	

A further distinction between periodic inspection and initial verification relates to the sequence of testing. *IEE Guidance Note 3* gives a different sequence for periodic testing to that which is given in *BS 7671* for new installation work. For comparison purposes, both sequences are listed on the opposite page.

Where, during the course of inspection or testing, a real and immediate danger is found to be present in an installation (from an exposed live part, for example), immediate action will be necessary before continuing. It is not sufficient simply to draw attention to the danger when submitting an inspection report. At the very least, the inspector must ensure that a person with responsibility for the safety of the installation is made aware of the danger that exists. An agreement should be made with this person as to the appropriate action to be taken to remove the source of danger (for example, by switching off and isolating the affected part of the installation until remedied), before continuing with the inspection or testing.

Contractors should note that the *Electricity at Work Regulations* effectively requires them to endeavour to make safe, before leaving site and with the agreement of the user or owner, any dangerous conditions found in installations. For example, where blanks are missing from a distribution board, suitable temporary barriers should be installed to protect persons from direct contact with live parts.

COMPILING THE FORM

Parts of the NICEIC Periodic Inspection Report form are identical or similar to parts of the NICEIC Electrical Installation Certificate and the NICEIC Domestic Electrical Installation Certificate, the preparation of which is described in Chapter 2 and Chapter 3, respectively. Where each of these parts is reached in this chapter, the reader is referred to Chapter 2 in order to avoid repetition.

SECTION A: DETAILS OF THE CLIENT

Client/Address

Insert details as described in Chapter 2 under 'Details of Client'.

Inspection, Testing, Certification and Reporting

This report is not valid if the serial number has been defaced or altered PIR/ **XXXXXXX**

PERIODIC INSPECTION REPORT
FOR AN ELECTRICAL INSTALLATION

Issued by an Approved Contractor or Conforming Body enrolled with the National Inspection Council for Electrical Installation Contracting, Vintage House, 37 Albert Embankment, London SE1 7UJ.

Original (To the person ordering the work)

See reverse of this page for explanatory notes relating to NICEIC software endorsement

A. DETAILS OF THE CLIENT

Client:

Address:

B. PURPOSE OF THE REPORT
This Periodic Inspection Report must be used only for reporting on the condition of an existing installation.

Purpose for which this report is required:

C. DETAILS OF THE INSTALLATION

Occupier:

Address:

Postcode:

Description of premises: Domestic Commercial Industrial

Other: (Please state)

Estimated age of the electrical installation: years

Evidence of alterations or additions   If yes, estimated age years

Date of previous inspection:

Electrical Installation Certificate No or previous Periodic Inspection Report No:

Records of installation available: Records held by:

Agreed limitations, if any, on the inspection and testing:

This inspection has been carried out in accordance with BS 7671: 1992 (IEE Wiring Regulations), as amended. Cables concealed within trunking and conduits, or cables and conduits concealed under floors, in inaccessible roof spaces and generally within the fabric of the building or underground, have not been visually inspected.

E. DECLARATION

I/We, being the person(s) responsible for the inspection and testing of the electrical installation (as indicated by my/our signatures below), particulars of which are described above (see C), having exercised reasonable skill and care when carrying out the inspection and testing, hereby declare that the information in this report, including the observations (see F) and the attached schedules (see H), provides an accurate assessment of the condition of the electrical installation taking into account the stated extent of the installation and the limitations of the inspection and testing (see D). I/We further declare that in my/our judgement, the said installation was overall in ✤ condition (see G) at the time the inspection was carried out, and that it should be further inspected as recommended (see I).

✤ (Insert 'a satisfactory' or 'an unsatisfactory', as appropriate)

INSPECTION, TESTING AND ASSESSMENT BY:

Signature:

Name: (CAPITALS)

Position:

Date:

REPORT REVIEWED AND CONFIRMED BY: † See note below

Signature:

Name: (CAPITALS)

(Registered Qualified Supervisor for the Approved Contractor at J)

Date:

Page 1 of

† This Periodic Inspection Report should be reviewed and confirmed by the registered Qualified Supervisor for the Approved Contractor responsible for issuing the Report.

This form is based on the model shown in Appendix 6 of BS 7671: 1992 as amended 1997
Published by the National Inspection Council for Electrical Installation Contracting © Copyright NICEIC (Jan 2000)

Please see the 'Notes for Recipients' on the reverse of this page.

106

SECTION B: PURPOSE OF THE REPORT

Purpose for which this report is required

It is important, for all concerned, that a clear statement of the purpose of the report is made in this part of the form. A report may be required, for example, in connection with a proposed house purchase or for an application for an entertainments licence. Alternatively, a report may be required on the condition of an installation following a fire or flood. A report may be required at the end of the period recommended in the original Electrical Installation Certificate or Domestic Electrical Installation Certificate, or a previous Periodic Inspection Report, or as an assessment of the condition of the installation in relation to current standards.

SECTION C: DETAILS OF THE INSTALLATION

Occupier

The occupier may or may not be the client as described in Section A. An appropriate name or title should be obtained and recorded in this box.

Address and postcode

The complete postal address is necessary, sufficient to clearly identify the premises containing the installation concerned.

Description of premises

'Domestic', 'Commercial' or 'Industrial' should be ticked as appropriate or, if the type of premises does not fall within any of these categories (such as a construction site), a brief description should be entered in the box marked 'Other'.

Estimated age of the installation

This may be determined from information relating to the original installation. Otherwise a reasonable estimate should be made by other means, such as the appearance of the installed equipment.

Evidence of alterations or additions

If there is clear evidence as to whether or not alterations or additions have been made to the installation since the previous inspection (or from new, if no subsequent information is available), this should be indicated by stating 'Yes' or 'No' in the box as appropriate.

NICEIC

This report is not valid if the serial number has been defaced or altered

PIR/ XXXXXXX

PERIODIC INSPECTION REPORT
FOR AN ELECTRICAL INSTALLATION

Issued by an Approved Contractor or Conforming Body enrolled with the National Inspection Council for Electrical Installation Contracting, Vintage House, 37 Albert Embankment, London SE1 7UJ.

See reverse of this page for explanatory notes relating to NICEIC software endorsement

Original (To the person ordering the work)

A. DETAILS OF THE CLIENT

Client:

Address:

B. PURPOSE OF THE REPORT
This Periodic Inspection Report must be used only for reporting on the condition of an existing installation.

Purpose for which this report is required:

C. DETAILS OF THE INSTALLATION

| | Domestic | Commercial | Industrial |

Occupier:

Description of premises:

Address:

Other: (Please state)

Estimated age of the electrical installation: years

Postcode:

Evidence of alterations or additions

If yes, estimated age years

Date of previous inspection:

Electrical Installation Certificate No or previous Periodic Inspection Report No:

Records of installation available:

Records held by:

Agreed limitations, if any, on the inspection and testing:

This inspection has been carried out in accordance with BS 7671: 1992 (IEE Wiring Regulations), as amended. Cables concealed within trunking and conduits, or cables and conduits concealed under floors, in inaccessible roof spaces and generally within the fabric of the building or underground, have not been visually inspected.

E. DECLARATION

I/We, being the person(s) responsible for the inspection and testing of the electrical installation (as indicated by my/our signatures below), particulars of which are described above (see C), having exercised reasonable skill and care when carrying out the inspection and testing, hereby declare that the information in this report, including the observations (see F) and the attached schedules (see H), provides an accurate assessment of the condition of the electrical installation taking into account the stated extent of the installation and the limitations of the inspection and testing (see D). **I/We further declare that in my/our judgement, the said installation was overall in** ✤ **condition (see G) at the time the inspection was carried out, and that it should be further inspected as recommended (see I).**

✤ *(Insert 'a satisfactory' or 'an unsatisfactory', as appropriate)*

INSPECTION, TESTING AND ASSESSMENT BY:

REPORT REVIEWED AND CONFIRMED BY: † *See note below*

Signature:

Signature:

Name: (CAPITALS)

Name: (CAPITALS)

(Registered Qualified Supervisor for the Approved Contractor at J)

Position:

Date:

Date:

Page 1 of

† This Periodic Inspection Report should be reviewed and confirmed by the registered Qualified Supervisor for the Approved Contractor responsible for issuing the Report.

This form is based on the model shown in Appendix 6 of BS 7671: 1992 as amended 1997
Published by the National Inspection Council for Electrical Installation Contracting © Copyright NICEIC (Jan 2000)

Please see the 'Notes for Recipients' on the reverse of this page.

If there is clear evidence of alteration or addition either visually or by reference to subsequent certificates, its actual or estimated age should be given in the space available. Otherwise write 'N/A' (meaning 'Not Applicable').

Date of previous inspection

It may be possible to ascertain the date of the previous inspection from a Periodic Inspection Report or the original Installation Certificate or equivalent. Alternatively, the 'Periodic Inspection Notice' fixed at the origin of the installation (to comply with Regulation 514-09-01) should provide an indication of the previous inspection. If the information cannot be ascertained, then 'not known' should be entered.

Electrical Installation Certificate No or previous Periodic Inspection Report No

Where the inspector has details of the original Electrical Installation Certificate or Domestic Electrical Installation Certificate, or a subsequent Periodic Inspection Report for the particular installation, the serial number of the latest document should be recorded in the space provided. If the information cannot be ascertained, then 'not known' should be entered.

Records of installation available

This refers to the availability of records of original certification, previous periodic inspection(s), operating and maintenance manuals, as-installed drawings etc. Indicate 'yes' or 'no', as appropriate.

Records held by

If the previous indication was 'yes', give the name of the person, company or organisation that is holding the records of the previous inspection(s). If the previous indication was 'no', write 'N/A'.

SECTION D: EXTENT OF THE INSTALLATION AND LIMITATIONS OF THE INSPECTION AND TESTING

Extent of electrical installation covered by this report

Before commencing an inspection, it is essential to agree with the client the exact extent of the installation to be inspected, taking into account the requirements of any third party involved (such as a licensing authority). It might be necessary to make a cursory inspection of the installation and available inspection records before discussing with the client the scope of the detailed inspection required, and the available options.

NICEIC

This report is not valid if the serial number has been defaced or altered

PIR/ XXXXXXX

PERIODIC INSPECTION REPORT
FOR AN ELECTRICAL INSTALLATION

Issued by an Approved Contractor or Conforming Body enrolled with the National Inspection Council for Electrical Installation Contracting, Vintage House, 37 Albert Embankment, London SE1 7UJ.

See reverse of this page for explanatory notes relating to NICEIC software endorsement

Original (To the person ordering the work)

A. DETAILS OF THE CLIENT

Client:

Address:

B. PURPOSE OF THE REPORT   This Periodic Inspection Report must be used only for reporting on the condition of an existing installation.

Purpose for which this report is required:

C. DETAILS OF THE INSTALLATION

Occupier:

Description of premises: Domestic   Commercial   Industrial

Address:

Other: (Please state)

Estimated age of the electrical installation: years

Postcode:

Evidence of alterations or additions   If yes, estimated age   years

Date of previous inspection:

Electrical Installation Certificate No or previous Periodic Inspection Report No:

D. EXTENT OF THE INSTALLATION AND LIMITATIONS OF THE INSPECTION AND TESTING

Extent of the electrical installation covered by this report:

Agreed limitations, if any, on the inspection and testing:

This inspection has been carried out in accordance with BS 7671: 1992 (IEE Wiring Regulations), as amended. Cables concealed within trunking and conduits, or cables and conduits concealed under floors, in inaccessible roof spaces and generally within the fabric of the building or underground, have not been visually inspected.

I/We, being the person(s) responsible for the inspection and testing of the electrical installation (as indicated by my/our signatures below), particulars of which are described above (see C), having exercised reasonable skill and care when carrying out the inspection and testing, hereby declare that the information in this report, including the observations (see F) and the attached schedules (see H), provides an accurate assessment of the condition of the electrical installation taking into account the stated extent of the installation and the limitations of the inspection and testing (see D). I/We further declare that in my/our judgement, the said installation was overall in ✦ _____ condition (see G) at the time the inspection was carried out, and that it should be further inspected as recommended (see I).

✦ (Insert '**a satisfactory**' or '**an unsatisfactory**', as appropriate)

INSPECTION, TESTING AND ASSESSMENT BY:

Signature:

Name: (CAPITALS)

Position:

Date:

REPORT REVIEWED AND CONFIRMED BY: † See note below

Signature:

Name: (CAPITALS)

(Registered Qualified Supervisor for the Approved Contractor at J)

Date:

Page 1 of

† This Periodic Inspection Report should be reviewed and confirmed by the registered Qualified Supervisor for the Approved Contractor responsible for issuing the Report.

This form is based on the model shown in Appendix 6 of BS 7671: 1992 as amended 1997
Published by the National Inspection Council for Electrical Installation Contracting © Copyright NICEIC (Jan 2000)

Please see the 'Notes for Recipients' on the reverse of this page.

The agreed extent of the installation covered by the inspection should be fully recorded in this part of Section D. The extent of the installation covered by the report might be the whole of the installation in the premises described in Section C, or it might be only a part of that installation. The part could, for example, be a particular area (such as the second floor only) or particular circuits (such as all lighting circuits). The extent could also be indicated by exclusions from the whole electrical installation, such as all high bay lighting or the fire alarm system. The wording will need to be tailored exactly to each inspection requirement and, in some instances, a continuation sheet (appropriately numbered, and identified here and in Section H) may be necessary.

Agreed limitations, if any, on the inspection and testing

The limitations imposed on the inspection and testing should also be explained to, and agreed with, the client before work commences. Section D of the NICEIC Periodic Inspection Report form includes standard text which states that concealed cables in certain parts of the building's structure will not be visually inspected. All other limitations should also be agreed before commencing the inspection and be clearly recorded on the report.

A periodic inspection must include a thorough visual inspection of all the electrical equipment that comprises the extent of the installation described in the previous section, and which is not concealed or inaccessible. However, in most cases, it will be appropriate to apply a sampling process to the detailed inspection of the internal condition of equipment, the condition of joints and terminations etc. It may also be appropriate to apply a sampling process to the inspection and testing of the installation.

If sampling is intended, this should be made clear to the client, and the initial degree of sampling for both the inspection and the testing elements should be agreed before the work commences. *IEE Guidance Note 3* provides recommendations for the minimum degree of sampling for both inspection and testing. Should the actual degree of sampling carried out be different from that agreed initially (because, for example, increased sampling was necessary due to deficiencies being found during the initial sampling), the actual degree should be recorded on the report. Details of the sampling technique and identification of the equipment selected for inspection and testing on a sampling basis should also be fully identified, on an additional page if necessary.

The contractor should not set unnecessary limitations on the inspection. The aim should be to produce a description that is both fair to the client and reasonable for the contractor. It should be remembered that the greater the limitations applying to a report, the less its value.

Inspection, Testing, Certification and Reporting

PIR/ XXXXXXX

PERIODIC INSPECTION REPORT
FOR AN ELECTRICAL INSTALLATION

Issued by an Approved Contractor or Conforming Body enrolled with the National Inspection Council for Electrical Installation Contracting, Vintage House, 37 Albert Embankment, London SE1 7UJ.

Original (To the person ordering the work)

See reverse of this page for explanatory notes relating to NICEIC software endorsement

A. DETAILS OF THE CLIENT

Client:

Address:

B. PURPOSE OF THE REPORT
This Periodic Inspection Report must be used only for reporting on the condition of an existing installation.

Purpose for which this report is required:

C. DETAILS OF THE INSTALLATION

Occupier:

Address:

Postcode:

Date of previous inspection:

Records of installation available:

Records held by:

Description of premises: Domestic   Commercial   Industrial

Other: (Please state)

Estimated age of the electrical installation: years

Evidence of alterations or additions   If yes, estimated age   years

Electrical Installation Certificate No or previous Periodic Inspection Report No:

D. EXTENT OF THE INSTALLATION AND LIMITATIONS OF THE INSPECTION AND TESTING

Extent of the electrical installation covered by this report:

Agreed limitations, if any, on the inspection and testing:

This inspection has been carried out in accordance with BS 7671: 1992 (IEE Wiring Regulations), as amended. Cables concealed within trunking and conduits, or cables and conduits concealed under floors, in inaccessible roof spaces and generally within the fabric of the building or underground, have not been visually inspected.

E. DECLARATION

I/We, being the person(s) responsible for the inspection and testing of the electrical installation (as indicated by my/our signatures below), particulars of which are described above (see C), having exercised reasonable skill and care when carrying out the inspection and testing, hereby declare that the information in this report, including the observations (see F) and the attached schedules (see H), provides an accurate assessment of the condition of the electrical installation taking into account the stated extent of the installation and the limitations of the inspection and testing (see D). **I/We further declare that in my/our judgement, the said installation was overall in** ✤ **condition (see G) at the time the inspection was carried out, and that it should be further inspected as recommended (see I).**

✤ (Insert 'a satisfactory' or 'an unsatisfactory', as appropriate)

INSPECTION, TESTING AND ASSESSMENT BY:

Signature:

Name: (CAPITALS)

Position:

Date:

REPORT REVIEWED AND CONFIRMED BY: † See note below

Signature:

Name: (CAPITALS)

(Registered Qualified Supervisor for the Approved Contractor at J)

Date:

Page 1 of

† This Periodic Inspection Report should be reviewed and confirmed by the registered Qualified Supervisor for the Approved Contractor responsible for issuing the Report.

This form is based on the model shown in Appendix 6 of BS 7671: 1992 as amended 1997
Published by the National Inspection Council for Electrical Installation Contracting © Copyright NICEIC (Jan 2000)

Please see the 'Notes for Recipients' on the reverse of this page.

Liabilities arising from inadequate description:

It should be appreciated that failure to clearly describe and record the extent of the installation and the agreed limitations of the inspection and testing could involve the contractor in unforeseen liabilities at a later date. For example, if a contractor failed to clearly record that all high bay lighting had been excluded from a periodic inspection report, the contractor might be held responsible if an electrical fault in that part of the installation later resulted in danger or damage.

If there is insufficient space in Section D to provide full details of the extent of the work to which the inspection report refers, the details should be provided on an additional page attached to the report, and reference to the additional page should be included in the relevant part(s) of the box.

SECTION E: DECLARATION

Section E is a declaration that the report gives an accurate assessment of the condition of the installation within the extent and limitations specified at Section D. The declaration should be signed by the person who carried out the inspection, and then by a person who is competent to review the contents of the report to check that it has been correctly compiled. In the case of an NICEIC Periodic Inspection Report, the reviewer should be the Approved Contractor's registered Qualified Supervisor*.

Where the Qualified Supervisor has personally conducted the inspection and testing, and has compiled the report, that person should sign in both places. In an organisation where a substantial quantity of periodic inspection reports need to be signed on a regular basis, it may be acceptable for the reviewer to delegate the countersigning of the inspection reports to a person of equivalent competence and responsibility, and to verify on a sampling basis that reports are being correctly compiled. However, in such cases the responsibility for the contents of all issued Periodic Inspection Reports remains with the Qualified Supervisor.

The data-entry box for the overall condition of the installation must be completed by inserting either 'a satisfactory' or 'an unsatisfactory'. This declaration of the condition of the installation should reiterate the detailed entry given in Section G: Summary of the Inspection, which itself should summarise the observations and recommendations made in Section F.

---

*   *'Qualified Supervisor' replaces the term 'Qualifying Manager' to align with the requirements of the new industry assessment scheme expected to be introduced under the auspices of BS EN 45011: 1998: General requirements for bodies operating product certification systems, and with the corresponding changes in the NICEIC Rules Relating to Enrolment.*

F. OBSERVATIONS AND RECOMMENDATIONS FOR ACTIONS TO BE TAKEN

Referring to the attached schedules of inspection and test results, and subject to the limitations at D:

There are no items adversely affecting electrical safety.

or

The following observations and recommendations are made.

Item No		Code †
1		

Original (To the person orde

Note: If necessary, continue on additional pages(s), which must be identified by the Periodic Inspection Report serial number and page number(s).

† Where observations are made, the inspector will have entered one of the following codes against each observation to indicate the action (if any) recommended:-

1. 'requires urgent attention' or   2. 'requires improvement' or
3. 'requires further investigation' or   4. 'does not comply with BS 7671: 1992 (as amended)'

Please see the reverse of this page for guidance regarding the recommendations.

Urgent remedial work recommended for Items:

Corrective action(s) recommended for Items:

G. SUMMARY OF THE INSPECTION

General condition of the installation:

Note: If necessary, continue on additional page(s), which must be identified by the Periodic Inspection Report serial number and page number(s).

Date(s) of the inspection:

Overall assessment of the installation:

(Entry should read either 'Satisfactory' or 'Unsatisfactory')

Page 2 of

This form is based on the model shown in Appendix 6 of BS 7671: 1992 as amended 1997
Published by the National Inspection Council for Electrical Installation Contracting © Copyright NICEIC (Jan 2000)

Please see the 'Guidance for Recipients on the Recommendation Codes' on the reverse of this page.

SECTION F: OBSERVATIONS AND RECOMMENDATIONS FOR ACTIONS TO BE TAKEN

Any observations and recommendations relating to the installation should, after due consideration, be provided in this section of the report.

The observations and recommendations should take due account of the results of the inspection and testing, and be based on the requirements of the issue of *BS 7671* current at the time of the inspection (**not** the requirements of an earlier standard current at the time the installation was constructed).

This section includes two small unshaded data-entry boxes at the top, one of which must be completed with a positive entry such as a 'Yes' or a '✔' to indicate that 'there are no items adversely affecting electrical safety' or, alternatively, 'the following observations and recommendations are made'. The data-entry box which does not attract a positive entry should be completed by recording 'N/A', meaning 'Not Applicable'.

In cases where observations and recommendations are appropriate, these are to be itemised and given a numerical reference under the left-hand column headed 'Item No'. The observation(s) must be provided in the wide centre column in an accurate, succinct and easily-understandable manner. Each observation must be attributed with a Recommendation Code 1, 2, 3 or 4, which is to be recorded in the right-hand column.

> Where an existing or potential danger is observed that may put the safety of those using the installation at risk, Recommendation Code 1 must be used.

If it is determined by inspection and testing that any item requires 'urgent attention', 'improvement' or 'further investigation', the data-entry box near the top of section, identified by the words 'The following observations and recommendation are made', should be completed by inserting a positive entry such as a 'Yes' or a '✔'.

Where Recommendation Codes 1, 2 or 3 have not been attributed to any of the items of observations, a positive entry (such as a 'Yes' or a '✔') should be given in the data-entry box identified by the words 'there are no items adversely affecting safety'. In these circumstances the two data-entry boxes at the bottom of the Section should each be completed by inserting the word 'None'.

GUIDANCE FOR RECIPIENTS ON THE RECOMMENDATION CODES

Recommendation Code 1

Where an observation has been given a Recommendation Code 1 (*requires urgent attention*), the safety of those using the installation may be at risk.

The person responsible for the maintenance of the installation is advised to take action without delay to remedy the observed deficiency in the installation, or to take other appropriate action (such as switching off and isolating the affected part(s) of the installation) to remove the potential danger. The NICEIC Approved Contractor issuing this report will be able to provide further advice.

It is important to note that the recommendations given at Section I *Next Inspection* of this report for the maximum interval until the next inspection, is **conditional** upon all items which have been given a Recommendation Code 1 being remedied without delay.

Recommendation Code 2

Recommendation Code 2 (*requires improvement*) indicates that, whilst the safety of those using the installation may not be at immediate risk, remedial action should be taken as soon as possible to improve the safety of the installation to the level provided by the national electrical safety standard BS 7671. The NICEIC Approved Contractor issuing this report will be able to provide further advice. Items which have been attributed Recommendation Code 2 should be remedied as soon as possible (see Section F).

Recommendation Code 3

Where an observation has been given a Recommendation Code 3 (*requires further investigation*), the inspection has revealed an apparent deficiency which could not, due to the extent or limitations of this inspection, be fully identified. Items which have been attributed Recommendation Code 3 should be investigated as soon as possible (see Section F).

The person responsible for the maintenance of the installation is advised to arrange for the NICEIC Approved Contractor issuing this report (or other competent person) to undertake further examination of the installation to determine the nature and extent of the apparent deficiency.

Recommendation Code 4

Recommendation Code 4 [*does not comply with BS 7671: 1992 (as amended)*] will have been given to observed non-compliance(s) with the current national electrical safety standard which do not warrant one of the other Recommendation Codes. It is not intended to imply that the electrical installation inspected is unsafe, but careful consideration should be given to the benefits of improving these aspects of the installation. The NICEIC Approved Contractor issuing this report will be able to provide further advice.

Guidance for recipients provided on the reverse of NICEIC Periodic Inspection Reports.

Where a positive entry has been given in the data-entry box identified by the words 'The following observations and recommendation are made', the data-entry boxes at the bottom of the Section are also required to be completed as follows:

- Items that have attracted a Recommendation Code 1, which indicates that urgent remedial work is required, must be prioritised by inserting the Item number(s) into the box identified with the words 'Urgent remedial work recommended for Items:'. Where no items have been attributed with a Recommendation Code 1 the entry must read 'None'.

- Items that have been attributed Recommendation Code 2 or 3, indicating 'requires improvement' and 'requires further investigation', respectively, must be entered into the data-entry box identified with the words 'Corrective actions(s) recommended for Items:'

Where a Recommendation Code 1 (requires urgent attention) is given, the client is to be advised immediately, in writing, that urgent work is necessary to remedy the deficiency to satisfy the duties imposed on the contractor and others by the *Electricity at Work Regulations 1989.* If the space on the form is insufficient, additional numbered pages are to be provided as necessary. Additional pages should also be identified by the unique Periodic Inspection Report number.

Observations:

Each recorded observation should describe a specific defect or omission in the electrical installation. The observation should detail what the situation is, and not what is considered necessary to put it right. For example, 'excessive external damage to the enclosure of Distribution Board F2/7' would be appropriate, whereas 'Distribution Board F2/7 to be replaced' would not. Remember that this is a factual report on the condition of an installation, not a proposal for remedial work.

Only observations that can be supported by one or more regulations in the current issue of *BS 7671* should be recorded. The particular regulation number(s) need not be entered on the report (unless specifically required by the client), but should serve to remind the inspector that it is only compliance with *BS 7671* that is to be considered. Observations based solely on personal preference or 'custom and practice' should not be included. It may be acceptable for example, under certain conditions, to have a pendant luminaire in a bathroom (see Regulation 601-11-01). The acceptability of such features should be properly considered and not automatically reported on adversely.

Each observation should be written in a manner that will be understood by the client. Comments should be clear and unambiguous, but the use of technical terms should be avoided unless it is known that the recipient is an electrical engineer or electrician, for example.

Typical defects which the NICEIC considers would warrant a Code 1 recommendation:

- Access to exposed live parts.

- Absence of means of earthing for the installation.

- A water service pipe being used as the means of earthing for the installation.

- Absence of circuit protective conductors for circuits supplying items of Class I equipment.

- Incorrect polarity.

- Absence of main equipotential bonding.

- Absence of local supplementary bonding where required (for example in bathrooms, shower rooms, swimming pools).

- Excessive earth fault loop impedance values.

- Absence of adequate protective devices (e.g. residual current devices for TT installations).

- Overloaded circuits.

- Circuits with ineffective overcurrent protection.

- Insufficient insulation resistance.

This list is not exhaustive and there are other defects that could be considered by the inspector to warrant urgent attention.

Recommendations:

Each observation must be given an appropriate recommendation code, selected from the standard codes numbered 1, 2, 3 and 4, as listed in the note near the foot of Section F. Each code has a particular meaning, as follows:

Recommendation Codes:

Code 1.	Requires urgent attention.
Code 2.	Requires improvement.
Code 3.	Requires further investigation.
Code 4.	Does not comply with *BS 7671*: 1992 (as amended).

One, and only one, of the standard recommendation codes should be attributed to each observation. If more than one recommendation could be applied to an observation, only the most onerous recommendation should be made (Code 1 being the most onerous). In general terms, the recommendation codes should be used as follows:

- **Code 1 (Requires urgent attention)** - Indicates that potential danger exists, requiring urgent remedial action. Examples of items that the NICEIC considers would warrant a Code 1 recommendation are given in the Table opposite.

- **Code 2 (Requires improvement)** - Indicates that the observed deficiency requires improvement to bring the standard up to a level that fully complies with *BS 7671*. For example, a Code 2 recommendation might be appropriate for a corroded metallic flexible conduit to a motor, or for a lighting circuit, wired in PVC insulated and sheathed cable not having a circuit protective conductor, but not at present supplying any item of Class I equipment. In this latter case a clear note should be included in Section G, 'Summary of Inspection', to the effect that the circuit is unsuitable for Class I equipment.

- **Code 3 (Requires further investigation)** - Indicates that the inspector was unable to come to a conclusion about this aspect of the installation, or that the observation was outside the agreed purpose, extent or limitations of the inspection, but has come to the inspector's attention during the inspection and testing. For example, it might not have been possible to trace a particular circuit. Such a recommendation would usually be associated with an observation on an aspect of the installation that was not foreseen when the purpose and extent of the inspection, and any limitations upon it, were agreed with the client. It should again be remembered that the purpose of periodic inspection is not to carry out a fault-finding exercise, but to assess and report on the condition of the installation within the agreed extent and limitations of the inspection.

N|C EIC

This report is not valid if the serial number has been defaced or altered

PIR/ XXXXXXX

Original (To the person ordering the work)

F. OBSERVATIONS AND RECOMMENDATIONS FOR ACTIONS TO BE TAKEN

Referring to the attached schedules of inspection and test results, and subject to the limitations at D:

There are no items adversely affecting electrical safety.

or

The following observations and recommendations are made.

Item No		Code †
1		

Note: If necessary, continue on additional pages(s), which must be identified by the Periodic Inspection Report serial number and page number(s).

† Where observations are made, the inspector will have entered one of the following codes against each observation to indicate the action (if any) recommended:-

1. 'requires urgent attention' or   2. 'requires improvement' or

G. SUMMARY OF THE INSPECTION

General condition of the installation:

Note: If necessary, continue on additional page(s), which must be identified by the Periodic Inspection Report serial number and page number(s).

Date(s) of the inspection:

Overall assessment of the installation:

(Entry should read either 'Satisfactory' or 'Unsatisfactory')

Page 2 of

This form is based on the model shown in Appendix 6 of BS 7671: 1992 as amended 1997
Published by the National Inspection Council for Electrical Installation Contracting © Copyright NICEIC (Jan 2000)

Please see the 'Guidance for Recipients on the Recommendation Codes' on the reverse of this page.

● Code 4 (Does not comply with *BS 7671: 1992* [as amended]) - Indicates that certain items have been identified as not complying with the requirements of *BS 7671*, but that the users of the installation are not in any danger as a result. For example, sleeving coloured green only may have been used on the termination of the circuit protective conductors at socket-outlets, or some minor detail might be missing from the installation schedule at a distribution board.

It is entirely a matter for the competent person conducting the inspection to decide on the recommendation code to be attributed to a given observation. The person's own judgement as a competent person should not be unduly influenced by the client. Remember that the persons signing the report are fully responsible for its content.

Further explanation of each recommendation is given on the reverse of NICEIC Periodic Inspection Report forms for the benefit of recipients, as previously indicated.

SECTION G: SUMMARY OF THE INSPECTION

The summary should adequately describe the overall condition of the installation, taking into account the specific observations made. It is essential to provide a clear summary of the condition of the installation having considered, for example:

● The adequacy of the earthing and bonding.

● The suitability of the switchgear and control gear.

● The type(s) of wiring system, and its condition.

● The serviceability of equipment, including accessories.

● The presence of adequate identification and notices.

● The extent of any wear and tear, damage or other deterioration.

● Changes in use of the building which have led to, or might lead to, deficiencies in the installation.

Minimal descriptions such as 'poor', and superficial statements such as 'Recommend a rewire' are considered unacceptable by the NICEIC as they do not indicate the true condition of the installation. It will, however, often be necessary or appropriate to explain the implications of a Periodic Inspection Report in a covering letter, for the benefit of recipients who require additional advice and guidance about their installation. For example, where an installation has deteriorated or been damaged to such an extent that its safe serviceable life can reasonably be considered to be at an end, a recommendation for renewal should be made in a covering letter, giving adequate supporting reasons. Reference to the covering letter should be made in the report.

Inspection, Testing, Certification and Reporting

n|c EIC

This report is not valid if the serial number has been defaced or altered   **PIR/ XXXXXXX**

Original (To the person ordering the work)

H. SCHEDULES AND ADDITIONAL PAGES

Schedule of Items Inspected and Schedules of Items Tested: Page No 4

Additional pages, including additional source(s) data sheets: Page No(s)

Schedule of Circuit Details for the Installation: Page No(s) 5

Schedule of Test Results for the Installation: Page No(s) 6

The pages identified here form an essential part of this report. The report is valid only if accompanied by all the schedules and additional pages identified above.

I. NEXT INSPECTION

I/We recommend that this installation is further inspected and tested after an interval of not more than

(Enter interval in terms of years, months or weeks, as appropriate)

provided that any items at F which have been attributed a Recommendation Code 1 (*requires urgent attention*) are remedied without delay. Items which have been attributed a Recommendation Code 2 or 3 should be actioned as soon as practicable (see F).

Address:

Telephone number:

Fax number:

Postcode:

n|c EIC Enrolment number: (Essential information)

Branch number: (if applicable)

K. SUPPLY CHARACTERISTICS AND EARTHING ARRANGEMENTS   *Tick boxes and enter details, as appropriate*

✧ System Type(s)	✧ Number and Type of Live Conductors			Nature of Supply Parameters			✧ Characteristics of Primary Supply Overcurrent Protective Device(s)	
TN-S	a.c.		d.c.	Nominal voltage(s), $U^{(1)}$	V $U_o^{(1)}$	V		
TN-C-S	1-phase (2 wire)	1-phase (3 wire)	2 pole	Nominal frequency, $f^{(1)}$	Hz	Notes: (1) by enquiry	BS(EN)	
TN-C	2-phase (3 wire)		3-pole	Prospective fault current, $I_{pf}^{(2)(3)}$	kA	(2) by enquiry or by measurement	Type	
TT	3-phase (3 wire)	3-phase (4 wire)	other	External earth fault loop impedance, $Z_e^{(2)(4)}$	Ω	(3) where more than one supply, record the higher or highest values	Nominal current rating	A
IT	Other	Please state		Number of supplies		(4) by measurement	Short-circuit capacity	kA

L. PARTICULARS OF INSTALLATION AT THE ORIGIN   *Tick boxes and enter details, as appropriate*

✧ Means of Earthing		Details of Installation Earth Electrode (where applicable)		
Supplier's facility:		Type: (eg rods, tape etc)	Location:	
Installation earth electrode:		Electrode resistance, R_A:	(Ω)	Method of measurement:

✧ Main Switch or Circuit-Breaker		Maximum Demand (Load):	A per phase	Method of Protection against Indirect Contact:	
(applicable only where an RCD is suitable and is used as a main circuit-breaker)					
Type: BS(EN)	Voltage rating	V	Main Protective Conductors		
No of Poles	Current rating, I_n	A	Earthing conductor	Main equipotential bonding conductors	Bonding of extraneous-conductive-parts (✓)
Supply conductors: material	RCD operating current, $I_{Δn}^*$	mA	Conductor material	Conductor material	Water service / Gas service
Supply conductors: csa	mm²	RCD operating time (at $I_{Δn}$)*	ms	Conductor csa	mm²	Conductor csa	mm²	Oil service / Structural steel
			Continuity check	(✓)	Continuity check	(✓)	Lightning protection / Other incoming service(s)

✧ *Where a number of sources are available to supply the installation, and where the data given for the primary source may differ from other sources, a separate sheet must be provided which identifies the relevant information relating to each additional source.*

Page 3 of

This form is based on the model shown in Appendix 6 of BS 7671: 1992 as amended 1997
Published by the National Inspection Council for Electrical Installation Contracting © Copyright NICEIC (Jan 2000)

The date (or dates if the inspection and testing were undertaken over a number of days) should be entered in the left-hand box at the bottom of this section. This is intended to be the date(s) when the inspection and testing were actually carried out, rather than when the report was issued.

At the bottom right-hand side of Section G of the report, a box is provided for the overall assessment of the condition of the installation to be given. After due consideration, the overall assessment should be given as either 'satisfactory' or 'unsatisfactory'. It would not be reasonable to indicate a 'satisfactory' assessment if any observation in the report had been given a 'Code 1' recommendation.

Where the space provided for the description of the general condition of the installation is inadequate for the purpose and it is necessary to continue the description on an additional page(s), the page number(s) of the additional page(s) should be recorded in this space, as indicated on the form.

SECTION H: SCHEDULES AND ADDITIONAL PAGES

This section is provided for the purpose of recording additional data sheets, including all the schedules relating to the report. Additional pages may be required where there is a need to record supplementary information about, for example, particular aspects of the installation or its source(s), which may include data sheets relating to the characteristics of additional sources where appropriate. The page number(s) of such additional sheets should be recorded in the top right-hand data-entry box ('Pages Nos') provided for this purpose.

Where additional schedules are used for the circuit details and associated test results, the page numbers should be recorded in the two data-entry boxes (after the printed page numbers which represent the minimum number of pages for the two schedules) provided to record the existence of these additional schedules which form part of the report.

SECTION I: NEXT INSPECTION

The inspector should make an appropriate recommendation for the time interval to the next periodic inspection, as described in Chapter 2 of this book. The Table given earlier in this Chapter gives some guidance in terms of maximum intervals. The guidance may be used directly if it is appropriate to the age, condition and usage of the installation. However, the inspector may decide to recommend a shorter interval than the maximum recommended in the table, due to the particular circumstances relating to the installation. This is acceptable if the inspector can properly justify such a recommendation. For example, a shorter period may be considered appropriate if an installation is old or perhaps subject to more rapid deterioration than usual.

n|c eic

This report is not valid if the serial number has been defaced or altered

PIR/ XXXXXXX

Original (To the the person ordering the work)

H. SCHEDULES AND ADDITIONAL PAGES

Schedule of Items Inspected and Schedules of Items Tested: Page No 4		Additional pages, including additional source(s) data sheets:	Page No(s)
Schedule of Circuit Details for the Installation: Page No(s) 5		Schedule of Test Results for the Installation: Page No(s)	6

The pages identified here form an essential part of this report. The report is valid only if accompanied by all the schedules and additional pages identified above.

I. NEXT INSPECTION

I/We recommend that this installation is further inspected and tested after an interval of not more than

(Enter interval in terms of years, months or weeks, as appropriate)

J. DETAILS OF NICEIC APPROVED CONTRACTOR

Trading Title:

Address:

Telephone number:

Fax number:

n|c eic Enrolment number: (Essential information)

Postcode:

Branch number: (if applicable)

K. SUPPLY CHARACTERISTICS AND EARTHING ARRANGEMENTS   *Tick boxes and enter details, as appropriate*

✦ System Type(s)	✦ Number and Type of Live Conductors			Nature of Supply Parameters		✦ Characteristics of Primary Supply Overcurrent Protective Device(s)	
TN-S	a.c.		d.c.	Nominal $U^{(1)}$ voltage(s):	V $U_0^{(1)}$ V		
TN-C-S	1-phase (2 wire)	1-phase (3 wire)	2 pole	Nominal frequency, $f^{(1)}$	Hz	*Notes:* *(1) by enquiry*	BS(EN)
TN-C	2-phase (3 wire)		3-pole	Prospective fault current, $I_{pf}^{(2)(3)}$	kA	*(2) by enquiry or by measurement*	Type
TT	3-phase (3 wire)	3-phase (4 wire)	other	External earth fault loop impedance, $Z_e^{(3)(4)}$	Ω	*(3) where more than one supply, record the higher or highest values*	Nominal current rating A
IT	Other	*Please state*		Number of supplies		*(4) by measurement*	Short-circuit capacity kA

L. PARTICULARS OF INSTALLATION AT THE ORIGIN   *Tick boxes and enter details, as appropriate*

✦ Means of Earthing		Details of Installation Earth Electrode (where applicable)		
Supplier's facility:		Type: (eg rod(s), tape etc)	Location:	
Installation earth electrode:		Electrode resistance, R_A:	(Ω)	Method of measurement:

✦ Main Switch or Circuit-Breaker *(applicable only where an RCD is suitable and is used as a main circuit-breaker)*		Maximum Demand (Load):	A per phase	Method of Protection against Indirect Contact:	

Type: BS(EN)		Voltage rating	V	**Main Protective Conductors**			
				Earthing conductor	**Main equipotential bonding conductors**	**Bonding of extraneous-conductive-parts (✓)**	
No of Poles		Current rating, I_n	A	Conductor material	Conductor material	Water service	Gas service
Supply conductors material		RCD operating current, $I_{\Delta n}$	mA	Conductor csa mm²	Conductor csa mm²	Oil service	Structural steel
Supply conductors csa	mm²	RCD operating time (at $I_{\Delta n}$)	ms	Continuity check (✓)	Continuity check (✓)	Lightning protection	Other incoming service(s)

✦ Where a number of sources are available to supply the installation, and where the data given for the primary source may differ from other sources, a separate sheet must be provided which identifies the relevant information relating to each additional source.

SECTION J: DETAILS OF NICEIC APPROVED CONTRACTOR

On an NICEIC Periodic Inspection Report, the Approved Contractor's trading title as registered with the NICEIC (not necessarily the Contractor's name), postal address and NICEIC Enrolment Number and Branch Number, where applicable, should be given here. The recording of the Approved Contractor's telephone and fax numbers is optional.

SECTION K: SUPPLY CHARACTERISTICS AND EARTHING ARRANGEMENTS

'The supply characteristics and earthing arrangements' should be obtained and inserted generally as described in Chapter 2. Note, however, that the NICEIC Periodic Inspection Report form requires the value for the external earth fault loop impedance (Z_e) to be obtained by measurement only. The NICEIC considers that the determination of this parameter by calculation or enquiry is not appropriate for an existing installation, and these options are therefore not catered for on the form.

Where, due to operational considerations, it is truly impracticable to measure the external earth fault loop impedance (Z_e), it would be acceptable to obtain a value from up-to-date records of the installation, provided there can be no doubt that the intended means of earthing is present and suitable for the installation. Such records should be available from the person responsible for the safety of the installation. Such considerations could, for example, apply to a hospital installation, where it might be unacceptable for the supply to be disconnected to enable a measurement of Z_e to be made for the purpose of periodic inspection. In such an exceptional case, the words 'by enquiry' should be included adjacent to the value of Z_e so obtained, and reference to the operational restriction should be included in the 'limitations of the inspection' part of Section D (see page 85).

SECTION L: PARTICULARS OF THE INSTALLATION AT THE ORIGIN

'The particulars of the installation at the origin' should be ascertained and inserted in the data-entry boxes all as described in Chapter 2.

Inspection, Testing, Certification and Reporting

Schedules for the Periodic Inspection Report

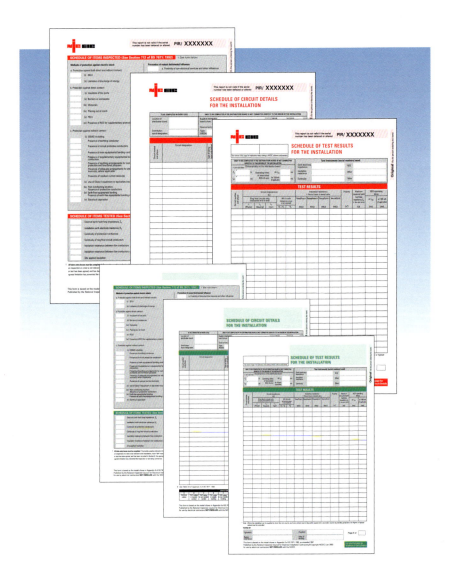

SCHEDULES

The remainder of the NICEIC Periodic Inspection Report comprises various schedules, which are common to those used for the NICEIC Electrical Installation Certificate, except that a 'Schedule of Additional Records', features only in the latter. Guidance on how to complete these schedules will be found in Chapter 6. Guidance on the process of inspection and testing necessary to complete the schedules is given in Chapters 7 and 8 respectively.

It should be noted that a Periodic Inspection Report must be issued with all its associated schedules, even where, due to operational or other restrictions, the work comprised a visual inspection only. The schedules should be completed to the fullest extent possible with regard to the limitations of the inspection. Where, due to the particular limitations placed on a periodic inspection, a schedule or parts of a schedule are not applicable, this should be indicated by 'N/A' or 'Not Applicable'. The NICEIC considers any NICEIC Periodic Inspection Report that is issued without appropriately compiled schedules to be incomplete, and therefore invalid.

Where required, continuation schedules are available separately from the NICEIC as indicated earlier in this Chapter.

Additional (continuation) schedules are required where the installation includes distribution ('sub-main') circuits - see Chapter 6.

Inspection, Testing, Certification and Reporting

Chapter

6

The Schedules

Inspection, Testing, Certification and Reporting

Schedules for the Electrical Installation Certificate and Periodic Inspection Report

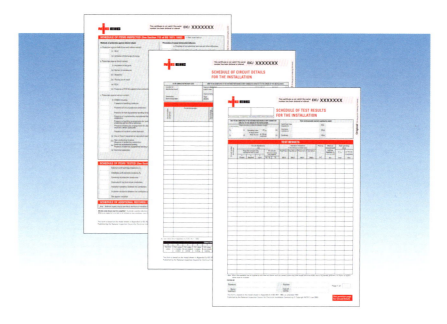

The schedules for the NICEIC Electrical Installation Certificate and Periodic Inspection Report are virtually identical.

Modified Schedule for the NICEIC Domestic Electrical Installation Certificate.

6 THE SCHEDULES

REQUIREMENTS

This chapter gives guidance on the preparation of the schedules which form part of both the NICEIC Electrical Installation Certificate and the NICEIC Periodic Inspection Report. The guidance is also generally applicable to the preparation of the schedules which form part of the NICEIC Domestic Electrical Installation Certificate. Guidance on the associated inspection and testing procedures is given in Chapters 7 and 8, respectively.

BS 7671 requires that a schedule of test results is issued with each Electrical Installation Certificate, Domestic Electrical Installation Certificate and Periodic Inspection Report, with the test schedule recording the results of the appropriate tests detailed in Regulations 713-02 to 713-12. The requirement for the recording of test results relates not only to final circuits but also to distribution circuits (or 'sub-mains' as they used to be called), where they form part of the installation.

Both the NICEIC Electrical Installation Certificate and the NICEIC Periodic Inspection Report incorporate the following schedules:

- Schedule of Items Inspected.
- Schedule of Items Tested.
- Schedule of Circuit Details for the Installation.
- Schedule of Test Results for the Installation.

Note: The Domestic Electrical Installation Certificate incorporates all these schedules in a modified form on the second page of the certificate.

The NICEIC Electrical Installation Certificate also makes provision (on Page 3) for a Schedule of Additional Records (which is not included in the NICEIC Periodic Inspection Report or the Domestic Electrical Installation Certificate).

The Schedule of Items Inspected and Schedule of Items Tested are used to identify the particular protective measures and other safety features that are applicable to the installation, and to confirm that these have been inspected and tested as required by *BS 7671*. The Schedule of Circuit Details and the Schedule of Test Results must be issued as a pair, giving details of each circuit of the installation. The former records information about the design, such as the composition of each circuit and the type and rating of each protective device, and the latter records the test results.

The four schedules common to the NICEIC versions of the Electrical Installation Certificate and the Periodic Inspection Report are virtually identical. The schedules incorporated on second page of the Domestic Electrical Installation Certificate are abridged versions of the schedules addressed in this chapter.

This certificate is not valid if the serial number has been defaced or altered

EIC/ XXXXXXX

SCHEDULE OF ITEMS INSPECTED (See Section 712 of BS 7671: 1992) † See note below

Methods of protection against electric shock

a. Protection against both direct and indirect contact:

(i) SELV

(ii) Limitation of discharge of energy

b. Protection against direct contact:

(i) Insulation of live parts

(ii) Barriers or enclosures

(iii) Obstacles

(iv) Placing out of reach

(v) PELV

(vi) Presence of RCD for supplementary protection

c. Protection against indirect contact:

(i) EEBAD including:

Presence of earthing conductor

Presence of circuit protective conductors

Presence of main equipotential bonding conductors

Presence of supplementary equipotential bonding conductors

Presence of earthing arrangements for combined protective and functional purposes

Presence of adequate arrangements for alternative source(s), where applicable

Presence of residual current device(s)

(ii) Use of Class II equipment or equivalent insulation

(iii) Non-conducting location: Absence of protective conductors

(iv) Earth-free equipotential bonding: Presence of earth-free equipotential bonding conductors

(v) Electrical separation

Prevention of mutual detrimental influence

a. Proximity of non-electrical services and other influences

b. Segregation of Band I and Band II circuits or Band II insulation used

c. Segregation of safety circuits

Identification

Presence of diagrams, instructions, circuit charts and similar information

Presence of danger notices and other warning notices

Labelling of protective devices, switches and terminals

Identification of conductors

Cables and Conductors

Routing of cables in prescribed zones or within mechanical protection

Connection of conductors

Erection methods

Selection of conductors for current carrying capacity and voltage drop

Presence of fire barriers, suitable seals and protection against thermal effects

General

Presence and correct location of appropriate devices for isolation and switching

Adequacy of access to switchgear and other equipment

Particular protective measures for special installations and locations

Connection of single-pole devices for protection or switching in phase conductors only

Correct connection of accessories and equipment

Presence of undervoltage protective devices

Choice and setting of protective and monitoring devices (for protection against indirect contact and/or overcurrent)

Selection of equipment and protective measures appropriate to external influences

Selection of appropriate functional switching devices

SCHEDULE OF ITEMS TESTED (See Section 713 of BS 7671: 1992) † See note below

External earth fault loop impedance, Z_e

Installation earth electrode resistance, R_A

Continuity of protective conductors

Continuity of ring final circuit conductors

Insulation resistance between live conductors

Insulation resistance between live conductors and earth

Site applied insulation

Protection by separation of circuits

Protection against direct contact by barrier or enclosure provided during erection

Insulation of non-conducting floors or walls

Polarity

Earth fault loop impedance, Z_s

Operation of residual current devices

Functional testing of assemblies

SCHEDULE OF ADDITIONAL RECORDS (See attached schedule)

Page No(s)

Note: Additional page(s), must be identified by the Electrical Installation Certificate serial number and page number(s).

† **All data-entry boxes must be completed.** To provide a positive indication that an inspection or a test has been carried out, insert either a 'Yes' or a '✓'. Where an inspection or a test is not relevant to the installation, insert 'N/A' meaning 'Not Applicable'.

Page 3 of

This form is based on the model shown in Appendix 6 of BS 7671: 1992, as amended 1997
Published by the National Inspection Council for Electrical Installation Contracting © Copyright NICEIC (Jan 2000)

SCHEDULE OF ITEMS INSPECTED AND SCHEDULE OF ITEMS TESTED

Each of these two schedules, shown opposite, have a number of unshaded boxes which must be completed. For schedules which form part of an Electrical Installation Certificate, the boxes must either be completed by inserting a 'Yes' or a '✔' to indicate that for items inspected or tested (as applicable) the task has been completed, or 'N/A' entered where an item is not applicable. To avoid doubt as to whether any particular inspection or test is applicable to the particular installation, no box should be left blank.

For schedules which are part of a Periodic Inspection Report, all unshaded boxes must be completed by inserting, or a 'Yes' or a '✔' to indicate that for items inspected or tested (as applicable) the task has been completed, or by entering 'N/A' meaning Not Applicable, where appropriate. Exceptionally, where a limitation on a particular inspection or test has been agreed with the client, and the reasons and technical justification for it have been recorded in Section D, the appropriate box(es) must be completed by inserting 'LIM', indicating that an agreed limitation has prevented the inspection or test being carried out. Again, to avoid doubt, no box should be left blank.

In the NICEIC's experience, the schedules issued as part of the Electrical Installation Certificate and Periodic Inspection Report are often completed incorrectly. For example, installations having a non-conducting location are uncommon since this protective measure is only used in special locations which are under effective supervision, yet it is surprising how often the related box is found to have been ticked on a Schedule of Items Inspected.

Other common errors include not ticking the boxes for items or systems which must be present if the installation work is being certified as compliant with *BS 7671*, or for items or systems that should have been considered when reporting on the condition of an existing installation in accordance with *BS 7671*.

SCHEDULE OF ADDITIONAL RECORDS

The Electrical Installation Certificate makes provision (on Page 3) for a Schedule of Additional Records (such a schedule is not included in the Periodic Inspection Report or in the Domestic Electrical Installation Certificate). This provision immediately follows the Schedule of Items Tested in the Electrical Installation Certificate. Where additional records (such as relevant drawings, specifications or previous inspection reports) form part of the certificate, the page number(s) of such records must be recorded in the data-entry box provided. Where no such records exist, the data-entry box should record 'None'.

All individual additional records must be identified by the Electrical Installation Certificate serial number and by an appropriate page number(s).

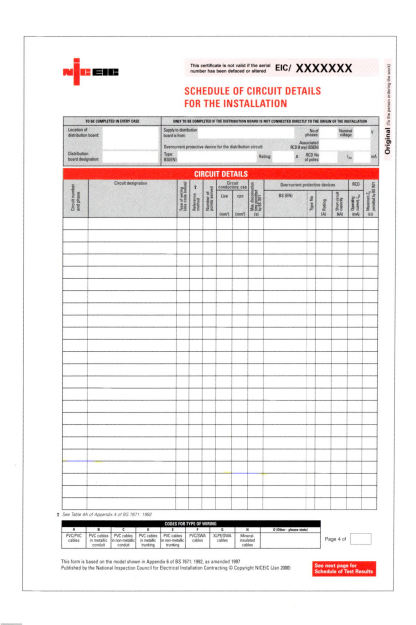

Chapter 6

SCHEDULE OF CIRCUIT DETAILS FOR THE INSTALLATION

A Schedule of Circuit Details must be compiled for each switchboard, distribution board, consumer unit, control panel and the like included in the installation covered by the Electrical Installation Completion Certificate or the Periodic Inspection Report. Reference to 'distribution board' in this chapter should to be taken to mean any of the above-mentioned items of switchgear, or composites of such items.

The schedule relates mainly to the design information for the installation, and should be completed in all possible respects prior to any testing being carried out on site. Each schedule includes essential details about the item of distribution equipment to which it relates, and about the outgoing circuits, whether they are final circuits or distribution circuits. It is important to record the presence of spare ways, identified in the sequence of the other circuits. In the case of new installation work, the schedule can be used to confirm that the installation complies with the intended design.

One Schedule of Circuit Details is provided as part of each NICEIC certificate or report. Where the installation has more than one distribution board, additional schedules are required in the form of continuation sheets. Each continuation sheet must be numbered in sequence with all other sheets making up the certificate or report. See the section entitled 'Continuation Schedules' at the end of this Chapter.

At the top of the schedule are a number of boxes. Two of these must be completed irrespective of whether the distribution board is connected directly to the origin:

- **Location of distribution board:** Clearly identify the location.

- **Distribution board designation:** Give the reference number or other identifying mark. This designation should be different from that of any other distribution board forming part of the installation.

The other boxes at the top of the schedule should be completed only if the distribution board is not connected directly to the origin of the installation (for example, not connected directly to the Public Electricity Supplier's low-voltage metering equipment):

- **Supply to distribution board is from:** Identify the equipment from which the distribution circuit feeding the distribution board is directly supplied. For example 'Switchboard No 1'.

- **Overcurrent protective device for the distribution circuit:** State the type and rating of the overcurrent device protecting the distribution circuit that feeds the distribution board. For example, **Type BS (EN):** 'BS 88', **Rating:** '63 A'.

- **No of phase(s):** Record the number of phases.

- **Nominal voltage:** Record the nominal voltage to earth, U_o, in units of volts.

NICEIC

This certificate is not valid if the serial number has been defaced or altered   **EIC/ XXXXXXX**

Original (To the person ordering the work)

SCHEDULE OF CIRCUIT DETAILS
FOR THE INSTALLATION

TO BE COMPLETED IN EVERY CASE	ONLY TO BE COMPLETED IF THE DISTRIBUTION BOARD IS NOT CONNECTED DIRECTLY TO THE ORIGIN OF THE INSTALLATION			
Location of distribution board:	Supply to distribution board is from:		No of phases:	Nominal voltage:	V
	Overcurrent protective device for the distribution circuit:		Associated RCD (if any) BS(EN)		
Distribution board designation:	Type: BS(EN)	Rating:	A	RCD No of poles:	I$_{\Delta n}$	mA

CIRCUIT DETAILS

Circuit number and phase	Circuit designation	Type of wiring (see code below) †	Reference method	Number of points served	Circuit conductors: csa		Max disconnection time permitted by BS 7671	Overcurrent protective devices					RCD	Maximum Z$_s$ permitted by BS 7671
					Live (mm²)	cpc (mm²)	(s)	BS (EN)	Type No	Rating (A)	Short-circuit capacity (kA)	Operating current I$_{\Delta n}$ (mA)		(Ω)

† *See Table 4A of Appendix 4 of BS 7671: 1992*

CODES FOR TYPE OF WIRING								
A	B	C	D	E	F	G	H	O (Other - please state)
PVC/PVC cables	PVC cables in metallic conduit	PVC cables in non-metallic conduit	PVC cables in metallic trunking	PVC cables in non-metallic trunking	PVC/SWA cables	XLPE/SWA cables	Mineral-insulated cables	

Page 4 of [　]

This form is based on the model shown in Appendix 6 of BS 7671: 1992, as amended 1997
Published by the National Inspection Council for Electrical Installation Contracting © Copyright NICEIC (Jan 2000)

See next page for Schedule of Test Results

- **Associated RCD (if any): BS (EN):** State the product specification number. For example, 'BS EN 61008'.

- **RCD No of poles:** Record the number of poles. For example, 2, 3 or 4.

- **(RCD) $I_{\Delta n}$:** Record the nominal operating (tripping) current of the RCD, in units of mA.

CIRCUIT DETAILS

The details of each outgoing circuit must be entered in the body of the schedule - one row being used for each circuit. The details to be entered under each column heading are as follows:

- **Circuit Number and Phase:** The circuit number and phase should be identified using a scheme which is compatible with that used at the actual distribution board on site. For example: 1R, 1Y, 1B etc, or 1(R), 2(Y), 3(B) etc.

- **Circuit Designation:** The entry should be as descriptive as possible so that it cannot be confused with any other circuit on the distribution board. A column of 'lighting' or 'power' as designations would not be acceptable.

- **Type of wiring:** Details of the type of wiring, such as PVC/PVC cables or PVC cables in metallic conduit etc, should be entered. A table at the bottom of the schedule lists a set of codes for use in this respect. Where Code O (other) is used, a note must be added in the space provided in the table to identify the specific type of wiring.

- **Reference method:** The reference method(s), as detailed in Appendix 4 of *BS 7671*, used for installation of the circuit must be recorded here.

- **Number of points served:** Record the number of items of current-using equipment or socket-outlets on the circuit.

- **Circuit conductors (mm²):** Completion of this column is straightforward for the live conductors (phase and neutral), as it is for the circuit protective conductor (cpc) if this is a single-core cable or a core of a multi-core cable. It is not so straightforward if the cpc is provided by the steel wire armouring of an armoured cable, the copper sheath of a mineral insulated cable, or by steel conduit or trunking. In such cases, the cpc should be indicated by its type, such as 'Arm', 'Con', 'Trun' and so on, rather than by its cross-sectional area. In specifying the use of armouring etc. as a circuit protective conductor, the designer of the circuit will have taken responsibility for ensuring its suitability for that purpose in principle.

- **Maximum disconnection time permitted by BS 7671 (s):** The maximum permitted disconnection time is determined by the requirements of *BS 7671*, having regard to factors such as the type and location of equipment supplied by

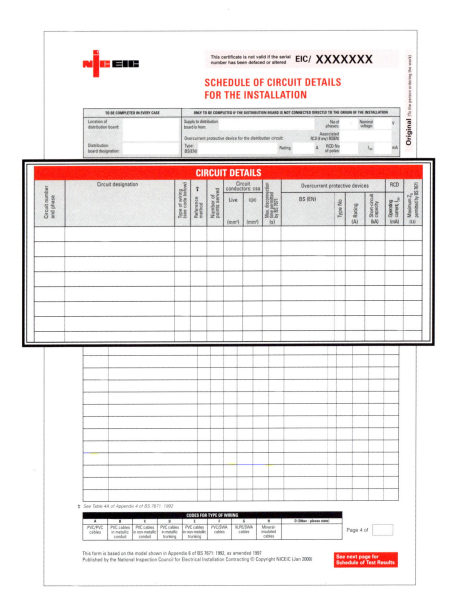

the circuit. For example, a socket-outlet circuit in a domestic dwelling would have a maximum permitted disconnection time of 0.4 seconds, but the maximum permitted disconnection time for a socket-outlet in agricultural premises might be 0.2 seconds. Circuits supplying fixed appliances will often have a maximum permitted disconnection time of 5 seconds, but this is reduced to 0.4 seconds in rooms containing a bath or shower basin, out of doors and so on. The designer should be consulted if there is any doubt.

- **Overcurrent protective devices:** There are four columns under this heading. The first two are for recording two pieces of information, namely the BS (EN) number and the type of the device. For example, the circuit overcurrent device may be a circuit-breaker to *BS EN 60898* and of Type B. The only exception is for a device that does not have 'Types', such as a *BS 3036* (rewirable) fuse. The nominal current or current setting of the protective device must be recorded (in units of Amps) in the third column under this heading, and the fourth column is for recording the short-circuit capacity of the device, in units of kA.

- **RCD: Operating current, $I_{\Delta n}$ (mA):** The figure to be inserted in this column is the rated residual operating (tripping) current, in mA, of any residual current device that may be in circuit. If an RCD has not been installed, then 'N/A' should be entered.

- **Maximum Z_s permitted by BS 7671 (Ω):** This column should record the maximum permitted values of Z_s, by reference to the limiting earth loop impedance values given Chapter 41 of *BS 7671*, not to other tabulated 'corrected' values used for comparison with measured values obtained at ambient temperature. Where there is an increased risk of electric shock, the limiting values given in Chapter 41 may be modified by the designer in accordance with the particular requirements specified in Part 6 of *BS 7671*.

Many distribution boards and consumer units do not have sufficient space to record all the information required by Regulations 514-08-01 and 514-09-01. For an installation for which the issue of a Domestic Electrical Installation Certificate has been appropriate, an additional copy of the completed schedules comprising the second page of the certificate is likely to provide most of the required information. For other simple installations, an additional copy of the 'Schedule of Circuit Details', which forms part of the NICEIC Electrical Installation Certificate, is likely to serve the same purpose. More complex installations will require more comprehensive information.

When an additional copy of the schedule is used to provide a record of the installation details, it should be protected from damage (for example by enclosing it in a clear plastic folder or encapsulation), and be fixed within or adjacent to the distribution board or consumer unit. The 'original' and 'duplicate' of the schedule should not be used for this purpose, but should be retained as part of the certificate by the client and contractor respectively.

This certificate is not valid if the serial number has been defaced or altered

EIC/ XXXXXXX

Original (To the person ordering the work)

SCHEDULE OF TEST RESULTS
FOR THE INSTALLATION

See reverse of page 1 for explanatory notes relating to NICEIC software endorsement

ONLY TO BE COMPLETED IF THE DISTRIBUTION BOARD IS NOT CONNECTED DIRECTLY TO THE ORIGIN OF THE INSTALLATION

✱ See note below — Characteristics at this distribution board

Z_s	Ω	Operating times of associated RCD (if any)	At $I_{\Delta n}$	ms
I_{pf}	kA		At 150mA (if applicable)	ms

Test instruments (serial numbers) used:

Earth fault loop impedance		RCD
Insulation resistance		Other
Continuity		Other

TEST RESULTS

Circuit number and phase	Circuit impedances (Ω)					Insulation resistance † Record lower or lowest value				Polarity	Maximum measured earth fault loop impedance, Z_s ✱ See note below	RCD operating times	
	Ring final circuits only (measured end to end)			All circuits (At least one column to be completed)		Phase/Phase †	Phase/Neutral †	Phase/Earth †	Neutral/Earth			at $I_{\Delta n}$	at 150 mA (if applicable)
	r_1 (Phase)	r_n (Neutral)	r_2 (cpc)	$R_1 + R_2$	R_2	(MΩ)	(MΩ)	(MΩ)	(MΩ)	(✓)	(Ω)	(ms)	(ms)

✱ Note: Where the installation can be supplied by more than one source, such as a primary source (eg public supply) and a secondary source (eg standby generator), the higher or highest values must be recorded.

TESTED BY

Signature:	Position:	Page 5 of
Name: (CAPITALS)	Date of testing:	

This form is based on the model shown in Appendix 6 of BS 7671: 1992, as amended 1997
Published by the National Inspection Council for Electrical Installation Contracting © Copyright NICEIC (Jan 2000)

See previous page for Circuit Details

SCHEDULE OF TEST RESULTS FOR THE INSTALLATION

In addition to the Schedule of Circuit Details previously discussed, a Schedule of Test Results should be compiled for each distribution board included in the installation covered by the Electrical Installation Certificate or the Periodic Inspection Report.

One such schedule is provided as part of each NICEIC certificate or report. Where the installation has more than one distribution board, additional schedules are required in the form of continuation sheets. There should be a corresponding Schedule of Test Results for each Schedule of Circuit Details. Each sheet should be numbered in sequence with all other sheets making up the certificate or report. See the section entitled 'Continuation Schedules' at the end of this chapter.

The boxes relating to characteristics at the distribution board at the top left-hand side of the schedule should be completed only if the distribution board is not connected directly to the origin of the installation.

- **Characteristics at this distribution board:** The values of earth fault loop impedance (Z_s) and maximum prospective fault current (I_{pf}), measured at the incoming terminals of the distribution board or control panel etc, should be recorded here, together with the respective operating times (ms) of an associated RCD (if fitted) when subject to a test current of $I_{\Delta n}$ and, where utilised for supplementary protection against direct contact, a test current 150 mA.

On the right-hand side at the top of the schedule is a panel where information concerning the test instruments used should be recorded:

- **Test instruments (serial numbers) used:** The serial number of each of the instruments used to obtain the test results recorded in this schedule should be given here. If any instrument does not have a serial number, a suitable number should be assigned and permanently marked on the instrument. Where a combined instrument such as an insulation/continuity test instrument is used to carry out more than one type of test, the serial number of that instrument should be given in the space corresponding to each of the relevant types of test. Where instruments have been used which are 'other' than those listed, delete the word 'other' and insert a brief description of each such instrument, followed by its serial number in the space provided.

NIC EIC

This certificate is not valid if the serial number has been defaced or altered **EIC/ XXXXXXX**

SCHEDULE OF TEST RESULTS
FOR THE INSTALLATION

See reverse of page 1 for explanatory notes relating to NICEIC software endorsement

ONLY TO BE COMPLETED IF THE DISTRIBUTION BOARD IS NOT CONNECTED DIRECTLY TO THE ORIGIN OF THE INSTALLATION

Characteristics at this distribution board *○ See note below*

Z_s *	Ω	Operating times of associated RCD (if any)	At $I_{\Delta n}$ ms
I_{pf} *	kA		At 150mA (if applicable) ms

Test instruments (serial numbers) used:

Earth fault loop impedance	RCD
Insulation resistance	Other
Continuity	Other

TEST RESULTS

Circuit number and phase	Circuit impedances (Ω)					Insulation resistance † Record lower or lowest value					Maximum measured earth fault loop impedance, Z_s *○ See note below* (Ω)	RCD operating times	
	Ring final circuits only (measured end to end)			All circuits (At least one column to be completed)		Phase/Phase †	Phase/Neutral †	Phase/Earth †	Neutral/Earth	larity (✓)		at $I_{\Delta n}$ (ms)	at 150 mA (if applicable) (ms)
	r_1 (Phase)	r_n (Neutral)	r_2 (cpc)	$R_1 + R_2$	R_2	(MΩ)	(MΩ)	(MΩ)	(MΩ)				

* Note: Where the installation can be supplied by more than one source, such as a primary source (eg public supply) and a secondary source (eg standby generator), the higher or highest values must be recorded.

TESTED BY

Signature:	Position:
Name: (CAPITALS)	Date of testing:

Page 5 of ____

This form is based on the model shown in Appendix 6 of BS 7671: 1992, as amended 1997
Published by the National Inspection Council for Electrical Installation Contracting © Copyright NICEIC (Jan 2000)

See previous page for Circuit Details

TEST RESULTS

The test results for each outgoing circuit must be entered in the body of the schedule, in the same rows and in the same order as the corresponding circuits appear on the Schedule of Circuit Details. This alignment is essential in order to enable the Schedule of Circuit Details and the Schedule of Test Results for each distribution board, to be read in conjunction with each other. The method for carrying out each test is described in Chapter 8. The information to be entered under each column heading is discussed below:

- **Circuit Number and Phase:** The circuit numbers and phase should be identical to those used on the corresponding Schedule of Circuit Details.

- **Circuit impedance: ring final circuits only (r_1, r_n, and r_2):** Where the circuit is ring final circuit, the resistance of each of the circuit's conductors, measured from end to end, should be recorded in the three columns provided, one each for the phase conductor ring (r_1), the conductor neutral ring (r_n) and the circuit protective conductor ring (r_2), in units of ohms. For circuits other than ring final circuits, 'N/A' meaning 'Not Applicable' must be entered for each of the three columns (r_1, r_n, and r_2).

- **Circuit impedance (R_1 + R_2):** Enter the measured resistance, in ohms, of the phase conductor and circuit protective conductor connected together as described for the measurement of (R_1 + R_2) in Chapter 8 (see 'continuity of protective conductors' or, for ring final circuits, 'continuity of ring final circuit conductors'). This test procedure also checks that polarity is correct.

- **Circuit impedance (R_2):** Where R_2 has been determined using the 'wander lead' method, or where for other reasons it is not practicable to record (R_1 + R_2), it is permissible to record R_2 values in this column.

Entry in at least one of the two columns, (R_1 + R_2) or (R_2), is necessary for all circuits, including ring final circuits. If, for technically justifiable reasons, data is not recorded in either of the two columns, 'N/A' meaning 'Not Applicable' must be entered.

- **Insulation resistance:** There are four columns under this heading. For a single-phase circuit, all but the first column are used ('N/A' being entered in the first column). For a 2-phase or 3-phase circuit, all columns are used, and the lower or lowest insulation resistance measured is recorded for phase/phase, phase/neutral and phase/earth. Enter measured values of insulation resistance in megohms. If the measured insulation resistance exceeds the maximum range of the test instrument, record the measurement as 'greater than the maximum range' (for example, > 200). Do not record the value as 'infinity'.

Inspection, Testing, Certification and Reporting

NICEIC

This certificate is not valid if the serial number has been defaced or altered

EIC/ **XXXXXXX**

Original (To the person ordering the work)

SCHEDULE OF TEST RESULTS
FOR THE INSTALLATION

See reverse of page 1 for explanatory notes relating to NICEIC software endorsement

ONLY TO BE COMPLETED IF THE DISTRIBUTION BOARD IS NOT CONNECTED DIRECTLY TO THE ORIGIN OF THE INSTALLATION

Characteristics at this distribution board

* See note below

Z_s ___ Ω Operating times of associated RCD (if any) At $I_{\Delta n}$ ___ ms

I_{pf} ___ kA At 150mA (if applicable) ___ ms

Test instruments (serial numbers) used:

Earth fault loop impedance ___ RCD

Insulation resistance ___ Other

Continuity ___ Other

TEST RESULTS

Circuit number and phase	Ring final (measured) r_1 (Phase)	(Ne	All circuits (At least one column to be completed) $R_1 + R_2$	R_2	Insulation resistance † *Record lower or lowest value*				Polarity	Maximum measured earth fault loop impedance, Z_s * See note below	RCD operating times	
					Phase/Phase † (MΩ)	Phases/Neutral † (MΩ)	Phases/Earth † (MΩ)	Neutral/Earth (MΩ)	(✓)	(Ω)	at $I_{\Delta n}$ (ms)	at 150 mA (if applicable) (ms)

TESTED BY

Signature: ___ Position: ___ Page 5 of ___

Name: (CAPITALS) ___ Date of testing: ___

This form is based on the model shown in Appendix 6 of BS 7671: 1992, as amended 1997
Published by the National Inspection Council for Electrical Installation Contracting © Copyright NICEIC (Jan 2000)

See previous page for Circuit Details

144

- **Polarity:** If polarity is found to be correct throughout the circuit by testing, a positive indication such as a tick (✔) should be entered. If polarity is found to be incorrect in any part of the circuit, a negative indication such as a cross (✘) should be entered. It is important to appreciate that an indication of incorrect polarity must never be included on a schedule of test results for new installation work; such a deficiency must always be corrected before the installation is put into service. Only in the case of a periodic inspection report can it ever be acceptable to issue a schedule of test results showing incorrect polarity. However, in the event of incorrect polarity constituting a real and immediate danger, this would require immediate action before continuing with the inspection and testing (as is the case with any defect which poses a real and immediate danger).

- **Maximum measured earth fault loop impedance, Z_s:** Enter the measured value of earth fault loop impedance at the point or accessory which is electrically most remote from the origin of the circuit.

- **RCD operating times:** There are two columns under this heading. The left-hand column is used to record the operating (tripping) time of any RCD which is in circuit when tested at its rated residual operating current. Where an RCD protects a number of circuits, the RCD (on no load) should be tested immediately downstream of the device, and the test currents and corresponding disconnection times may be recorded for the group of circuits so protected. Alternatively, the tests may be carried out on each circuit with the corresponding disconnection times recorded separately for each circuit.

  The right-hand column is applicable only if the RCD has a rated residual operating (tripping) current of 30 mA or less, and has been installed to provide supplementary protection against direct contact. This second test is carried out at a current of 150 mA. Test results are recorded in milliseconds (ms).

TESTED BY

The signature, name and position of the inspector, together with the date on which the tests were undertaken, should be entered at the foot of the page in the spaces provided for the purpose. Additionally, the total number of pages which constitute the certificate or report should be stated.

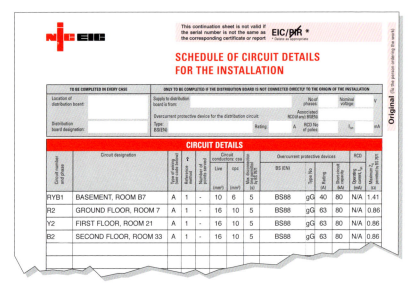

The serial number of the related certificate or report must be stated in the space provided on each schedule.

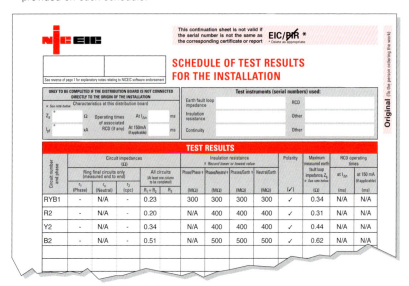

CONTINUATION SCHEDULES

As previously indicated, only one Schedule of Circuit Details and one Schedule of Test Results is supplied as part of each NICEIC Electrical Installation Certificate and Periodic Inspection Report.

Where the certificate or report relates to an installation having more than one distribution board or consumer unit, more than 24 circuits, or where the installation includes one or more distribution ('sub-main') circuits, additional schedules are required.

Sets of continuation schedules are available separately from the NICEIC. Only the version of the NICEIC continuation schedules intended to be used with the current versions of the NICEIC certificate and report forms should be used. Superseded versions of NICEIC continuation schedules should not be used.

The NICEIC version of the additional schedules should be given the same unique serial number as the other pages of the associated NICEIC Electrical Installation Certificate or Periodic Inspection Report, by first striking out the 'EIC' or 'PIR' prefix (as appropriate) at the top of the additional schedules and then adding the remainder of the serial number in the spaces allocated.

The page number for each additional schedule should be inserted at the foot of the schedule, together with the total number of pages comprising the certificate or report (eg page 6 of 7).

Due to the specific limitations placed on the use of the Domestic Electrical Installation Certificate, it would not be appropriate to issue continuation schedules with that certificate. If the nature of the installation work is such that the use of continuation schedules is necessary, an Electrical Installation Certificate should be issued for that work.

There is a tendency to overlook the need to record circuit details and test results for distribution circuits, where such circuits form part of an installation. For example, where a switchboard at the origin of an installation incorporates a number of fused-switches feeding distribution circuits to remote distribution boards, the information relating to those distribution circuits will need to be recorded on a Schedule of Circuit Details and a Schedule of Test Results. Such additional schedules are also required for the distribution circuit where a main switch at the origin of an installation feeds a single, remote distribution board.

An illustration of how the information relating to distribution circuits should be recorded is shown in the figure overleaf. The example is based on an installation having a switchboard at the origin having four outgoing ways feeding remote distribution boards. The Electrical Installation Certificate or Periodic Inspection Report must also include a separate Schedule of Circuit Details and Schedule of Test Results for each of the distribution boards, including their final circuits.

Inspection, Testing, Certification and Reporting

Chapter 7

INSPECTION

Inspection, Testing, Certification and Reporting

INSPECTION

PURPOSE OF INSPECTION

Before an installation is tested, it is necessary to carry out a careful and thorough inspection using the senses of touch, hearing and smell, as well as sight. Such inspection may reveal deficiencies that are unlikely to be detected by testing, such as exposed live parts, incomplete enclosures, damaged accessories or cable sheaths, unsuitable installation methods, overheating, absence of fire barriers etc.

The main purpose of inspection is to verify that the installed equipment and other associated materials:

- Comply with appropriate British, European or equivalent product standards.

- Have been correctly selected and erected.

- Are not visibly damaged, defective or deteriorated such that safety is impaired.

- Are suitable for the environment in which they are installed.

SAFETY

Wherever possible, inspection should be carried out when the installation is not energised, in accordance with the *Electricity at Work Regulations*. Where diagrams, schedules etc for the installation are not available, a degree of exploratory work may be necessary so that the inspection (and subsequent testing) can be carried out safely.

INSPECTION OF NEW INSTALLATION WORK

BS 7671 requires that every installation, and every alteration or addition to an installation, is inspected during erection and/or on completion before being put into service to verify, so far as is reasonably practicable, that the requirements of the standard have been met.

It is almost always necessary for some inspection work to be carried out during the process of constructing an installation, because some of the items requiring inspection are likely to be inaccessible on completion. For example, the condition of cables concealed in the fabric of a building cannot be inspected after completion of the building work. Inspection during the construction stage will also enable unsatisfactory work to be identified at an early stage, so that it can be remedied before becoming a serious and costly problem to the contractor.

N|C EIC

This certificate is not valid if the serial number has been defaced or altered

EIC/ XXXXXXX

Original (To the person ordering the work)

SCHEDULE OF ITEMS INSPECTED (See Section 712 of BS 7671: 1992) † See note below

Methods of protection against electric shock

a. Protection against both direct and indirect contact:

    (i) SELV

    (ii) Limitation of discharge of energy

b. Protection against direct contact:

    (i) Insulation of live parts

    (ii) Barriers or enclosures

    (iii) Obstacles

    (iv) Placing out of reach

    (v) PELV

    (vi) Presence of RCD for supplementary protection

c. Protection against indirect contact:

    (i) EEBAD including:

    Presence of earthing conductor

    Presence of circuit protective conductors

    Presence of main equipotential bonding conductors

    Presence of supplementary equipotential bonding conductors

    Presence of earthing arrangements for combined protective and functional purposes

    Presence of adequate arrangements for alternative source(s), where applicable

    Presence of residual current device(s)

    (ii) Use of Class II equipment or equivalent insulation

    (iii) Non-conducting location: Absence of protective conductors

    (iv) Earth-free equipotential bonding: Presence of earth-free equipotential bonding conductors

    (v) Electrical separation

Prevention of mutual detrimental influence

a. Proximity of non-electrical services and other influences

b. Segregation of Band I and Band II circuits or Band II insulation used

c. Segregation of safety circuits

Identification

Presence of diagrams, instructions, circuit charts and similar information

Presence of danger notices and other warning notices

Labelling of protective devices, switches and terminals

Identification of conductors

Cables and Conductors

Routing of cables in prescribed zones or within mechanical protection

Connection of conductors

Erection methods

Selection of conductors for current carrying capacity and voltage drop

Presence of fire barriers, suitable seals and protection against thermal effects

General

Presence and correct location of appropriate devices for isolation and switching

Adequacy of access to switchgear and other equipment

Particular protective measures for special installations and locations

Connection of single-pole devices for protection or switching in phase conductors only

Correct connection of accessories and equipment

Presence of undervoltage protective devices

Choice and setting of protective and monitoring devices (for protection against indirect contact and/or overcurrent)

Selection of equipment and protective measures appropriate to external influences

Selection of appropriate functional switching devices

SCHEDULE OF ITEMS TESTED (See Section 713 of BS 7671: 1992) † See note below

External earth fault loop impedance, Z_e

Installation earth electrode resistance, R_A

Continuity of protective conductors

Continuity of ring final circuit conductors

Insulation resistance between live conductors

Insulation resistance between live conductors and earth

Site applied insulation

Protection by separation of circuits

Protection against direct contact by barrier or enclosure provided during erection

Insulation of non-conducting floors or walls

Polarity

Earth fault loop impedance, Z_s

Operation of residual current devices

Functional testing of assemblies

SCHEDULE OF ADDITIONAL RECORDS (See attached schedule)

Note: Additional page(s), must be identified by the Electrical Installation Certificate serial number and page number(s).

Page No(s)

† **All data-entry boxes must be completed.** To provide a positive indication that an inspection or a test has been carried out, insert either a 'Yes' or a '✓'. Where an inspection or a test is not relevant to the installation, insert 'N/A' meaning 'Not Applicable'.

Page 3 of

This form is based on the model shown in Appendix 6 of BS 7671: 1992, as amended 1997

Published by the National Inspection Council for Electrical Installation Contracting © Copyright NICEIC (Jan 2000)

It is a requirement of *BS 7671* that the relevant design information is made available to those carrying out the inspection and testing of new installation work, irrespective of whether the design was carried out by the contractor or by another party such as a consulting engineer (see Regulation 711-01-02). Unless such information is made available at the appropriate time, it is unlikely that inspection and testing can be completed satisfactorily.

IN-SERVICE INSPECTION OF AN INSTALLATION

Electrical installations deteriorate with age, as well as with wear and tear. Every installation therefore needs to be inspected at appropriate intervals during its service life to check that its condition is such that it is safe to remain in service, and is likely to remain safe at least until the next inspection is due. Periodic inspection also enables the need for any remedial work to be identified and drawn to the attention of those responsible for the safety of the installation.

THE SEQUENCE OF INSPECTIONS

The sequence of the guidance for the various inspections in the remainder of this Chapter closely follows that of the items of inspection embodied in the NICEIC Schedule of Items Inspected, which forms part of both the NICEIC's Electrical Installation Certificate and the Periodic Inspection Report. The Domestic Electrical Installation Certificate also incorporates a modified form of this schedule. It is recognised that it may not be appropriate to follow such a sequence in undertaking the inspections.

METHODS OF PROTECTION AGAINST ELECTRIC SHOCK

Terminology

- **Direct Contact** means coming into contact with parts of electrical equipment which are intended to be live in normal use. This includes touching uninsulated live conductors, and coming into contact with live parts because an enclosure is damaged or inadequate.

- **Indirect Contact** means coming into contact with exposed-conductive-parts of electrical equipment which have become live as the result of an earth fault.

A visual check is required to verify that the measures intended by the designer to protect against direct and indirect contact have been implemented. *BS 7671* separates the measures into three groups:

- Measures of protection against both direct and indirect contact.

- Measures of protection against direct contact.

- Measures of protection against indirect contact.

Separated Extra-Low Voltage (SELV)

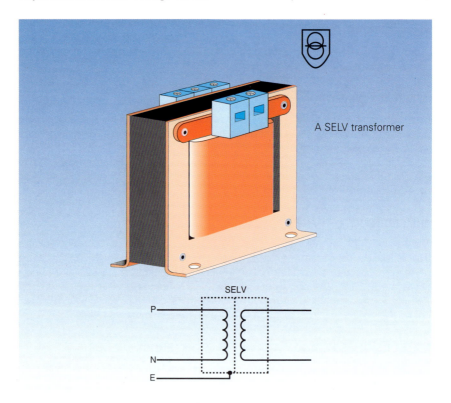

A SELV transformer

SELV

A safety isolating transformer to *BS EN 60742 (BS 3535)*.

Measures of protection against both direct and indirect contact

There are two measures that can provide protection against both direct contact and indirect contact:

SELV (Separated Extra-Low Voltage):

Inspection of this protective measure includes checking that the system source is suitable. Several types of source for SELV systems are recognised by *BS 7671*, but the most common is a safety isolating transformer complying with *BS EN 60742 (BS 3535:* Part 1), having no connection to earth on the secondary extra-low voltage winding. Such transformers are designed to provide an output of no more than 50 V.

Other types of source for SELV system recognised in *BS 7671* are:

- A motor-driven generator in which the windings provide electrical separation equivalent to that of a *BS EN 60742 (BS 3535:* Part 1) safety isolating transformer.

- An electro-chemical source, such as a battery, or another source independent of a higher voltage circuit, such as an engine-driven generator.

- Certain electronic devices which meet the constraints laid down in *BS 7671* and which are deemed to be suitable for the purpose.

A socket-outlet on a SELV circuit must be dimensionally incompatible with those used for other circuits. Additionally, live parts of a SELV circuit, including the transformer secondary winding and socket-outlets, must have no connection to earth, or to any protective conductor forming part of another electrical system. The visual inspection should therefore confirm that a PELV system (ie an Extra-low Voltage (ELV) system which is not electrically separated from earth but does otherwise comply with the requirements for SELV system) is not installed in a location where the design specifically requires a SELV system.

Limitation of discharge of energy:

This protective measure is one that is applied to an individual item of current-using equipment, such as an electric fence controller to *BS EN 61011* or *BS EN 61011-1*. Such items of equipment incorporate a means of limiting the output to safe levels of current and energy. Direct contact with an electric fence may be painful, but the energy contained in the discharge is not of a level likely to cause a dangerous physiological effect to the body of a person or livestock.

Inspection must determine that the equipment is marked as compliant with the relevant product standard, and therefore satisfies the requirements for this protective measure. It must also be verified that such circuits are separated from other circuits (as for SELV) and that there is not more than one fence controller connected to any fence system. Simultaneous operation of fence controllers could result in a discharge greater than a safe level, thereby negating the protective measure.

Measures of protection against direct contact

Insulation of live parts:

An inspection of the parts that will be live in service is required to verify that all the necessary insulation is present, has not been damaged during construction or use, and is in a sound and serviceable condition.

Barriers or enclosures:

An understanding of *BS EN 60529: Specification for degrees of protection provided by enclosures (IP code)*, and of the particular requirements of *BS 7671*, is required for the inspection of barriers and enclosures. For the purpose of this book, however, the basic requirements are that:

- Barriers or enclosures containing live parts must provide a degree of protection of at least IP2X or IPXXB (with certain exceptions).
- The horizontal top surface of a barrier or enclosure which is readily accessible must provide a degree of protection of at least IP4X.

    - IP2X means that the enclosure will not permit the insertion of any object 12.5 mm or more in diameter.
    - IPXXB means no contact with live parts is possible with a British Standard test finger having a diameter of 12 mm and 80 mm long.
    - IP4X means that at no point on the surface must the insertion of a wire or object greater than 1 mm thick be possible.

The letter 'X' in these IP Codes indicates that no particular degree of protection against ingress of liquids has been specified. It does not necessarily mean that no such protection is provided.

Generally, it should be possible to confirm by thorough visual inspection alone that barriers and enclosures provide the necessary minimum degree of protection against the ingress of solid objects. For example, it should be verified that all unused entries in enclosures have been closed, and that no blanks are missing from spare ways in distribution boards.

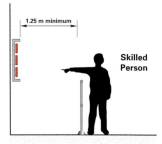

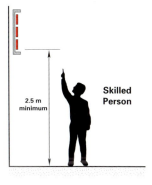

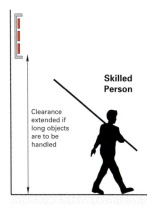

Obstacles:

An obstacle is something like a set of railings, the *BS 7671* definition being 'a part preventing unintentional contact with live parts but not preventing deliberate contact'. An obstacle may be removed without the use of a tool or key, but it should be secured so as to prevent unintentional removal. This protective measure may be used only in locations to which access is restricted to skilled persons, or to instructed persons under the direct supervision of a skilled person. The presence of warning notices associated with such restriction must be verified.

Placing out of reach:

This protective measure is intended to prevent unintentional contact with a bare live part by placing it out of reach. The measure may be used only in locations to which access is restricted to skilled persons, or to instructed persons under the direct supervision of a skilled person. The presence of such restriction must be verified.

Inspection of this protective measure will include verifying that the minimum clearance distances satisfy the definition of 'arms reach' given in Part 2 of *BS 7671*. Where long or bulky objects, such as ladders, are likely to be handled in the location and there is a risk of contact with live parts, the designer should have allowed for suitably increased clearance distances, and such increased distances must be verified by inspection.

Protective Extra-Low Voltage (PELV)

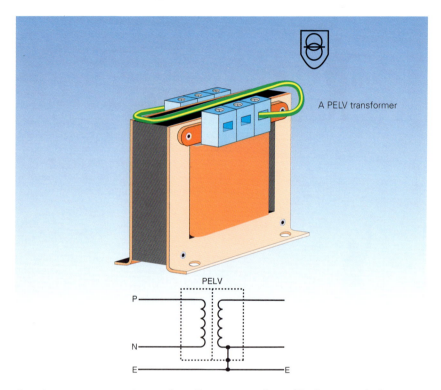

A PELV transformer

PELV

P

N

E ——————————————— E

Supplementary protection against direct contact by residual current devices

Each socket-outlet rated at 32 A or less which may reasonably be expected to supply portable equipment for use outdoors requires protection by a residual current device.

PELV (Protective Extra-Low Voltage):

A PELV system is similar to a SELV system, except that the extra-low voltage circuit is connected to earth at one point. This means that a PELV system will not be suitable in situations where a SELV system has been specified, such as for certain zones of swimming pools. The visual inspection should confirm that PELV systems are not installed in locations where the design requires SELV.

Supplementary protection against direct contact by residual current device

The inspector should check that supplementary protection against direct contact has been provided by a residual current device (RCD) for the following types of point or circuit, as required by *BS 7671*.

- Socket-outlets rated at 32 A or less which may reasonably be expected to supply portable equipment for use outdoors (except where SELV, electrical separation or automatic disconnection and reduced low voltage is incorporated in the circuit - see Regulation 471-16-01).

- Circuits supplying portable equipment for use outdoors, connected other than through a socket-outlet by means of a flexible cable or cord having a current-carrying capacity of 32 A or less (subject to the same exceptions as above).

- Designated points or circuits in the special installations or locations covered by Part 6 of *BS 7671* or, where intended by the designer, other installations or locations of increased shock risk.

An RCD must not have been provided as the sole means of protection against direct contact. One or more of the basic measures for protection against direct contact, referred to previously, must also have been provided.

Each device provided for supplementary protection against direct contact should be inspected to confirm that it has a rated residual operating current not exceeding 30 mA and that it meets a product standard which requires the device to operate under type-test conditions within 40 ms at a residual current of 150 mA. The number of points or circuits controlled by a single RCD should be such that the operation of that RCD would not result in unreasonable inconvenience (see Regulation 314-01-01).

Presence of earthing conductor

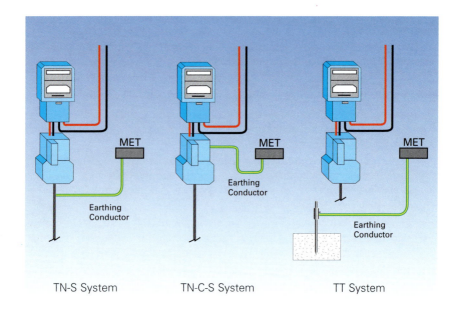

TN-S System   TN-C-S System   TT System

The earthing conductor connects the main earthing terminal (MET) of an installation to the means of earthing.

Chapter 7

Measures of protection against indirect contact

There are five basic measures for protection against indirect contact, at least one of which must be used for any installation.

Earthed Equipotential Bonding and Automatic Disconnection of supply (EEBAD), including:

Presence of earthing conductor

The earthing conductor is the protective conductor which connects the main earthing terminal of the installation to the means of earthing (eg the installation earth electrode for a TT system). The earthing conductor is a vitally important part of the earth fault loop, and it must be confirmed by inspection that the conductor is present and properly connected. The inspection must also confirm that the earthing conductor has a cross-sectional area of at least that required by *BS 7671*. Testing alone is not sufficient to determine that the earthing conductor has been correctly selected and installed.

The minimum cross-sectional area of the earthing conductor should have been determined by the designer in the same way as for any other protective conductor. This may have been by calculation, or by selection in accordance with Table 54G of *BS 7671*. Additionally, where the earthing conductor is buried in the ground, its cross-sectional area must not be less than that given in Table 54A of *BS 7671*.

If the cross-sectional area of the earthing conductor is less than that indicated in Table 54G (or Table 54A, if applicable), the inspector should consult the designer to confirm the adequacy of the earthing conductor arrangement or, for a periodic inspection, should make an appropriate observation and recommendation.

Presence of circuit protective conductors

A visual inspection must be made to check that circuit protective conductors, where required, have been correctly selected and installed. The designer may have determined the cross-sectional area of the circuit protective conductors either by calculation or by application of Table 54G of *BS 7671*. The chosen method of determination, together with other design information including the intended cross-sectional areas of each protective conductor, should be made available to the inspector.

Circuit protective conductors may be separate or incorporated in cables, or be formed by metallic cable sheath or armouring. Metallic enclosures such as conduit, trunking and so on may also be used provided that the design satisfies the requirements of Section 543 of *BS 7671*, in terms of its continuity, cross-sectional area and reliability.

Presence of main equipotential bonding conductors

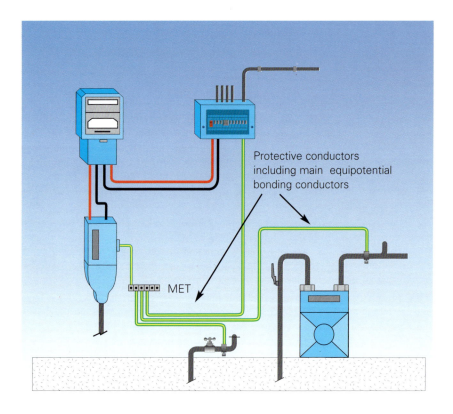

Protective conductors including main equipotential bonding conductors

MET

Presence of main equipotential bonding conductors

Main equipotential bonding conductors are bonding conductors connecting the main earthing terminal of the installation to extraneous-conductive-parts, which may include:

- Water service pipes.

- Gas service pipes.

- Other service pipes and ducting.

- Central heating and air conditioning systems.

- Exposed metallic structural parts of buildings.

- Lightning protection systems.

Where plumbing and/or wet central heating systems employ non-metallic pipework, it is difficult for the inspector to decide whether or not such pipework (or the liquid within it) constitutes an extraneous-conductive-part. For a new installation, cases of doubt should be referred to the installation designer for consideration and confirmation as to the requirement for main, and where applicable, supplementary bonding.

For the periodic inspection of an existing installation, the inspector will need to make an assessment based on all the relevant circumstances. *IEE Guidance Note 5: Protection Against Electric Shock* gives advice to help determine whether or not a conductive part is an extraneous-conductive-part.

Except where Protective Multiple Earthing (PME) conditions apply, the cross-sectional area of main bonding conductors must be not less than half the cross-sectional area required by *BS 7671* for the earthing conductor, with a minimum of 6 mm². The cross-sectional area need not be more than 25 mm² if the bonding conductor is of copper. If the bonding conductor is not of copper, its cross-sectional area must be such that its conductance is not less than that of the required size of copper conductor.

For an installation connected to a PME supply (a TN-C-S system), the cross-sectional area of the main bonding conductors must be not less than that given in Table 54H of *BS 7671*, see page 136. The data shown in this Table are **minimum** cross-sectional areas of main bonding conductors. The electricity supplier may require a larger cross-sectional area in some instances.

In the case of a service pipe, the main bonding conductor should have been connected as near as practicable to the point at which the particular service enters the premises.

The connection must be made on hard pipework, before any branch, on the consumer's side of any meter or insulating section. Where practicable, the connection should be made within 600 mm of the meter outlet. Where the meter is external, the bonding connection should be made at the point(s) of entry of the service into the building.

Inspection, Testing, Certification and Reporting

Minimum cross-sectional area (csa) of main equipotential bonding conductors for installations served by PME supplies

Copper equivalent csa of the supply neutral conductor	Minimum copper equivalent csa of the main equipotential bonding conductors
35 mm² or less	10 mm²
over 35 mm² up to 50 mm²	16 mm²
over 50 mm² up to 95 mm²	25 mm²
over 95 mm² up to 150 mm²	35 mm²
over 150 mm²	50 mm²

Notes:

The above data has been reproduced from Table 54H of *BS 7671* by kind permission of the IEE. This Table should be used as a guide only. The specific advice of the Public Electricity Supplier should always be sought if any doubt exists as to their requirements.

Where the inspector finds, in an existing installation, that the main bonding conductors have a cross-sectional area of say 6 mm², rather than (say) 10 mm² as required by *BS 7671*, the inspector may, having carefully considered all the circumstances, conclude that the deficiency does not pose a serious risk to the users. For example, where the installation has been modified by an addition or alteration, and where the existing main equipotential bonding satisfies the requirements of *BS 7671* except in terms of cross-sectional area and colour-coding, it may have been reasonable for the designer to have relied on that bonding as part of the protective measures for that addition or alteration. However, the deficiency must be recorded, together with an appropriate recommendation, on the Periodic Inspection Report, Electrical Installation Certificate, Domestic Electrical Installation Certificate or Minor Electrical Installation Works Certificate, as appropriate.

This dispensation is not applicable to a new installation, including a rewire. Neither is the dispensation applicable to an existing installation served by a PME supply which was commenced, or in certain cases has been worked upon, by the Public Electricity Supplier (PES) on or after 1 October 1988. Even where a PME supply was commenced before this date, careful consideration should be given to any decision to rely on existing main bonding conductors with cross-sectional areas less than required by Table 54H of *BS 7671*. There is always the possibility that main bonding conductors may have to carry network circulating current continuously or for long periods.

Chapter 7

Presence of supplementary equipotential bonding conductors

The requirement for supplementary equipotential bonding is generally applicable only to installations and locations where there is a perceived increased risk of electric shock. Some of these installations and locations are covered in Part 6 of *BS 7671*. There may be other circumstances where the designer has decided that an increased risk of electric shock exists which warrants the provision of supplementary equipotential bonding.

Supplementary equipotential bonding is required for most of the locations and installations covered by Part 6. The most common of these locations are rooms containing a bath or shower, and swimming pools. However, construction sites and restrictive conductive locations are also subject to requirements for supplementary bonding. It needs to be remembered that the particular requirements laid down in Part 6 of *BS 7671* supplement or modify the general requirements set out elsewhere in the Standard.

Where the circuit protective conductor in a short length of flexible cord has been used also to provide supplementary bonding to a fixed appliance as permitted by Regulation 547-03-05, it must be confirmed that a supplementary bonding conductor has been connected to the earthing terminal in the connection unit or other accessory for that appliance, in addition to the circuit protective conductor for that circuit.

The purpose of checking for the presence of supplementary equipotential bonding conductors in bathrooms and shower rooms is to confirm that simultaneously accessible exposed-conductive-parts and extraneous-conductive-parts have been connected together with suitably sized and correctly colour-coded (green-and-yellow) supplementary bonding conductors as required by *BS 7671*. Some of the supplementary bonding may be provided by extraneous-conductive-parts, subject to the requirements of Regulation 543-02-06. In other locations, the purpose is to check that supplementary bonding has been provided as intended by the designer.

Presence of earthing arrangements for combined protective and functional purposes

The installation design should indicate where provision has been made for conductors to serve as both functional and protective conductors. Where a protective conductor serves both purposes, the safety requirements relating to the protective role take precedence.

The installation design should show where conductors are used for combined functional and protective purposes. The inspection should confirm that such conductors are present, are of adequate cross-sectional area and are correctly identified in terms of colour-coding (green-and-yellow). The colour cream is reserved for conductors provided for functional earthing purposes only.

Inspection, Testing, Certification and Reporting

Adequate arrangements for alternative source(s):

Where a supply to the installation is available from more than one source, the inspection should identify that adequate arrangements in terms of protection for safety are present, including appropriate provision for isolation, manual and automatic switching, and any necessary interlocking. The particular requirements for generators including those used as parallel and standby sources are given in Section 551 of *BS 7671*.

Presence of residual current device(s):

The inspector should verify the presence of any residual current devices (RCDs) required by the designer for protection against indirect contact. *BS 7671* requires an RCD to be provided in any circuit where the earth fault loop impedance is too high for protection against indirect contact to be provided by other means. This is often the case with a TT system, for example. It should also be verified that every socket-outlet circuit in an installation forming part of a TT system is protected by an RCD in accordance with Regulation 471-08-06 (ie meets the requirement $Z_s I_{\Delta n} \leq 50$ V given in Regulation 413-02-16).

Use of Class II equipment or equivalent insulation:

Class II equipment should be identified by the Class II construction mark.

This protective measure is generally applicable only to individual items of equipment such as particular types of luminaire or space heater. The inspector should check that the measure is not being used as the sole means of protection against indirect contact for a whole installation or circuit unless this has been intended by the designer. In such a case, the designer must have been satisfied that the strict limitations placed upon such use by Regulation 471-09-03 of *BS 7671*, relating to effective supervision, will be applied throughout the lifetime of the installation.

It is to be expected that most equipment providing this protective measure will comprise type-tested items, marked to the appropriate standards. The inspector should check the markings on such equipment to see that it has double or reinforced insulation (Class II equipment), or total insulation in the case of any low-voltage switchgear and controlgear assemblies used to provide this type of protection (see *BS EN 60439*). A check should also be made that Class II protection has not been impaired by the method of installing the equipment.

BS 7671 also gives requirements relating to site-applied insulation. This includes supplementary insulation applied to electrical equipment having basic insulation only, and reinforced insulation applied to uninsulated live parts. The latter is recognised only where constructional features prevent the application of double insulation. Site-applied insulation must provide a degree of safety equivalent to the type-tested equipment referred to in the preceding paragraphs. Detailed inspection of site-applied insulation is beyond the scope of this book. Reference should be made to the designer and to the relevant British Standards for the criteria to be checked.

Whatever the type of equipment providing protection by Class II or equivalent insulation, the inspector should also check that the installation complies with the requirements of Regulations 413-03-03 to 413-03-09, and 471-09-01 to 471-09-04 which addresses the protective measure against indirect contact by Class II equipment or by equivalent insulation, and its application requirements, respectively.

Non-conducting location, and earth-free equipotential bonding:

These protective measures, although included in *BS 7671*, are not recognised in the Regulations for common use. They are intended to be applied only in special situations where effective supervision is provided, and they are prohibited in certain installations and locations of increased shock risk as covered in Part 6 of *BS 7671*. Owing to their specialised nature and comparative rarity of application, these protective measures are not included within the scope of this book.

Reference should be made to Regulations 413-04 and 471-10 of *BS 7671* for the main requirements relating to protection by non-conducting location, and Regulations 413-05, 471-11 and 514-13-02 for those relating to protection by earth-free equipotential bonding. Some guidance on the particular requirements for inspection and testing for these protective measures is given in *IEE Guidance Note 3*.

Electrical separation:

Electrical separation is a protective measure that may be applied to circuits operating at a voltage of up to 500 V. It should not be confused with extra-low voltage systems, SELV or PELV.

The inspector should check that both:

- the source of supply to the electrically-separated circuit complies with Regulation 413-06-02 (ie that the source is either an isolating transformer complying with *BS EN 60742 (BS 3535)* of which the secondary winding is not earthed, or that it is one of the other specific sources listed in that Regulation), and

- the electrically separated circuits comply with Regulation 413-06-03 (ie no live part is connected at any point to another circuit or to earth, either deliberately or unintentionally, as could occur due to a damaged cable; flexible cables and cords liable to mechanical damage are visible throughout their length; the wiring comprises either a separate wiring system (preferably) or one of the alternatives given in the Regulation (such as suitable cables in insulating conduit); overcurrent protection is provided; and electrical separation from other circuits is provided -see Chapter 8).

Inspection, Testing, Certification and Reporting

Use of electrical separation to supply an individual item of equipment

Where electrical separation is used to supply an individual item of equipment, the inspector should verify that no exposed-conductive-part of the separated circuit, such as any metal casing of an item of equipment fed by the circuit, is connected to either the protective conductor of the source, or to an exposed-conductive-part of another circuit.

Use of electrical separation to supply more than one item of equipment

The use of electrical separation to supply several items of equipment from a single source is recognised in the Regulations only for special locations under effective supervision, and where specified by a suitably qualified electrical engineer. Inspection of such an installation is beyond the scope of this book. The main requirements are covered by Regulations 413-06-05, 471-12-01 and 514-13-02.

PREVENTION OF MUTUAL DETRIMENTAL INFLUENCE

Proximity of non-electrical services and other influences

It should be verified that no destructive, damaging or harmful effects have occurred, are actually occurring, or are likely to occur, because of the proximity of any part of the electrical installation to non-electrical services or other influences. For example, it should be checked that any heat, steam, smoke, fumes or condensation is not likely to damage the electrical installation; or that electrolytic corrosion is not likely to occur between dissimilar metal parts of the electrical installation and the building structure or mechanical plant which are in contact with each other in damp situations, such as outdoors.

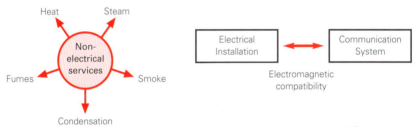

Examples of possible mutual detrimental influences

Where the equipment or wiring of an electrical installation is in the vicinity of a communication or control system (say), it should be checked that the relevant electromagnetic compatibility (EMC) requirements have been complied with, so that one system will not cause interference to the other. Regulations 515-02-01 and 515-02-02 require that equipment, when installed, will:

- Be able to withstand interference levels present in the vicinity.

- Not interfere with other equipment in the vicinity.

Where there is a lift, a check should be made to verify that the lift well has not been used as a route for cables which do not form part of the lift installation, as defined in *BS 5655*. This restriction on the use of the lift well is mainly due to considerations of fire risk and to the problems of gaining safe access for inspection, testing and maintenance of parts not related to the lift installation, rather than to the existence of mutual detrimental influences. This is, however, a convenient heading under which to include this check.

Inspection, Testing, Certification and Reporting

Segregation of Band I and Band II circuits or Band II insulation used

It is necessary to verify by inspection that Band I and Band II circuits are separated or segregated from each other either by containment in separate wiring systems or by adopting a method of insulating (either individually or collectively) the Band I conductors for the highest voltage present, according to Regulations 528-01-02 and 528-01-07.

Voltage Band I covers installations where protection against electric shock is provided under certain conditions by the value of voltage. It extends also to installations where the voltage is limited for operational reasons such as in telecommunication, signalling, bell control and alarm installations. Circuits operating at voltages not exceeding extra-low voltage (50 V ac or 120 V ripple-free dc) will normally fall within Band I.

Voltage Band II includes low-voltage supplies to household, commercial and industrial installations. Low voltage is defined as exceeding extra-low voltage but not exceeding 1000 V ac or 1500 V dc between conductors (which are also the upper limits on Voltage Band II), or 600 V ac or 900 V dc between conductors and Earth.

Segregation or separation may be achieved in a number of ways:

- By containment of a Band I circuit, insulated for its system voltage, in a separate conduit, trunking or ducting system.

- By containment of Band I and Band II circuits insulated for their respective system voltage in separate compartments of a common trunking or ducting system.

- By installation of Band I and Band II circuits on tray or ladder where physical separation is provided by a partition.

- Where Band I and Band II circuits are contained in a common wiring system, by using circuit conductors all having an insulation rating suitable for the highest voltage present.

- Where Band I and Band II circuits are contained in a multicore cable or cord, by having the cores for Band I circuits suitably insulated for the highest voltage present, either individually or collectively. Alternatively, the cores of the Band I circuits must be separated from the Band II circuits by an earthed metal screen having a current-carrying capacity not less than that of the largest core of the Band II circuits.

Where outlets or controls for Band I and Band II circuits are mounted in or on a common box, block or switchplate of wiring systems formed by conduit, trunking or ducting, verification by inspection is required to check that the necessary segregation is maintained by an effective partition and that, where the partition is metal, it is earthed.

Segregation of safety circuits

Fire alarm and emergency lighting circuits are required to be segregated from other circuits and from each other in accordance with *BS 5266* and *BS 5839*. This typically means separation by a distance of not less than 300 mm, by separate enclosures, by continuous partition(s) in a common channel or trunking (with separation being maintained at cross-overs and boxes etc), or by wiring such circuits in mineral insulated cable. Amendment No 3 to *BS 5839: Part 1: 1988 - Fire Detection and Alarm Systems in Buildings* - also recognises cables complying with *BS 7629* as a means of separating conductors carrying fire alarm power and signals from conductors used for other systems. The segregation requirements also apply to the mains supply to fire alarm and emergency lighting systems.

BS 5839: Part 1 and *BS 5266: Part 1* contain requirements for segregation of the wiring of fire alarm and centralised emergency lighting systems.

Though not directly related to segregation, it must be appreciated that *BS 5266* and *BS 5839* also contain specific requirements concerning the selection and erection of emergency lighting and fire alarm wiring systems. The requirement is to provide both prolonged operation during a fire, and protection against mechanical damage. The electromagnetic compatibility requirements of Regulations 515-02-01 and 512-02-02 of *BS 7671* must also be satisfied.

Inspection, Testing, Certification and Reporting

IDENTIFICATION

Presence of diagrams, instructions, circuit charts and similar information

The presence of the necessary diagrams, charts or tables, or an equivalent form of information for the work being inspected, should be verified. The form of information should be legible and durable, and should indicate all of the following:

- The type and composition of each circuit (that is, the points of utilisation served, number and size of conductors, type of wiring).

- The method used for compliance with Regulation 413-01-01 (that is, the method of protection against indirect contact).

- The information required by Regulation 413-02-04 (that is, the type of earthing arrangement for example TT, TN-S, TN-C-S, and the types and current ratings and/or rated residual currents of protective devices).

- The information necessary for the identification of each device performing the functions of protection, isolation and switching, and its location.

- Any circuit or equipment vulnerable to a typical test.

For an installation for which the issue of a Domestic Electrical Installation Certificate has been appropriate, an additional copy of the completed schedules comprising the second page of the certificate, fixed within or adjacent to the consumer unit, is likely to provide most of the above information necessary for compliance with *BS 7671*. For other simple installations, an additional copy of the 'Schedule of Circuit Details', which forms part of the NICEIC Electrical Installation Certificate, is likely to serve the same purpose. More complex installations will require more comprehensive information.

In all cases, each switchboard, distribution board and consumer unit should have a copy of the relevant installation details provided in or adjacent to it. The schedule should include identification of any circuits or equipment vulnerable to a typical test.

Presence of danger notices and other warning notices

The inspector should verify that all relevant warning notices required by *BS 7671* have been fitted in the appropriate locations, and that they are durable and clearly visible. The most commonly required notices are as follows:

- **Voltage warning (Regulation 514-10-01)**
  This warning is to be visible before gaining access to live parts in an enclosure where a voltage exceeding 250 V exists but would not normally be expected.

- ### Voltage warning (Regulation 514-10-01)

  If terminals or other fixed live parts between which there is a voltage exceeding 250 V are housed in separate enclosures or items of equipment, but close enough to each other to be reached simultaneously by a person, a notice should be secured in a position such that anyone, before gaining access to the live parts, is warned of the voltage that exists between them.

- ### Isolation (Regulation 514-11-01)

  Required at each position where there are live parts not capable of being isolated by a single device. The location of each isolating device must be indicated unless there is no possibility of confusion.

- ### Periodic inspection (Regulation 514-12-01)

  To be fitted on completion of installation work (including alterations and additions) and on each completion of periodic inspection, in a prominent position at or near the origin, giving the recommended date by which the installation should be reinspected.

  IMPORTANT

  This installation should be periodically inspected and tested and a report on its condition obtained, as prescribed in BS 7671 (formally the IEE Wiring Regulations for Electrical Installations) published by the Institution of Electrical Engineers.
  Date of last Inspection
  Recommended date of next inspection

- ### RCD (Regulation 514-12-02)

  Required at the origin of an installation which incorporates one or more residual current devices. A similar notice may also be fitted adjacent to an RCD, wherever located.

  This installation, or part of it, is protected by a device which automatically switches off the supply if an earth fault develops. Test quarterly by pressing the button marked 'T' or 'Test'. The device should switch off the supply and should then be switched on to restore the supply. If the device does not switch off the supply when the button is pressed, seek expert advice.

- ### Earthing and bonding (Regulation 514-13-01)

  A notice should be fixed near:

  - The connection of an earthing conductor to an installation earth electrode.

  - The connection of a bonding conductor to an extraneous-conductive-part.

  - The main earthing terminal, where it is separate from the main switchgear.

  SAFETY ELECTRICAL CONNECTION
  DO NOT REMOVE

- ### Earth-free local equipotential bonding and electrical separation (Regulation 514-13-02)

  Required in a prominent position adjacent to every point of access to a location where the protective measure 'earth-free local equipotential bonding' is applied, or to the points of access to a location where 'electrical separation' is used to supply several items of equipment. (These are special situations which are required to be kept under effective supervision).

Inspection, Testing, Certification and Reporting

- Others
  - At all points of access to areas reserved for skilled or instructed persons (Regulation 471-13-03).
  - Fireman's switch (Regulations 476-03-07 and 537-04-06).
  - Caravans (Section 608).

Labelling of protective devices, switches and terminals

The inspector should verify the presence of suitable labelling or marking of protective devices, switches, terminals and the connections of conductors (where necessary). Labels and markings should be legible and durable.

Protective devices:

Each fuse, circuit-breaker and residual current device should be checked to ensure that:

- It is arranged and identified so that the circuit protected may be easily recognised.
- Its nominal current is indicated on or adjacent to it.

For protective devices incorporated in a distribution board, the above details may be given on the chart within or adjacent to it, provided the devices themselves are identified and arranged to correspond with the chart.

Switches etc:

The inspector should verify that, except where there is no possibility of confusion, each item of switching apparatus for operation, regulation or other control, has a label or other suitable means of identification to indicate its purpose and the installation, or circuit, it controls.

This isolator disconnects the supply to **No 1 BUCKET ELEVATOR**	**No 1 BUCKET ELEVATOR** The isolator for this machine is situated in Plant Room B

Terminals and connections of conductors:

The inspector should verify that, so far as is reasonably practicable, the wiring is arranged and/or marked so that it may be readily identified for the purposes of inspection, testing, alteration and repair of the installation.

Taking a distribution board or consumer unit as an example, the wiring within the equipment should be identified by arrangement. This usually means connection of the neutral and cpc of each circuit to the terminals which relate specifically to the associated outgoing fuse or circuit-breaker, by sequence or terminal marking. Where the arrangement does not identify the wiring sufficiently well for the purposes of the above regulation, the terminations of the conductors should be marked with codings such as 1, 2, 3' or 1R, 1Y, 1B' etc, as appropriate. Similarly, conductors connected to a main earthing terminal which is separate from the main switchgear should be identified by arrangement or marking.

Identification of conductors

The correct identification of the conductors of cables, and of any bare conductors, in terms of their polarity or protective function, should be verified. This is in addition to the requirements for identification of terminals and connections referred to in the preceding paragraphs.

Cables used as fixed wiring:

Single-core non-flexible cables, and cores of non-flexible cables should be identifiable at their terminations, and preferably throughout their length, by one of the methods listed in Regulation 514-06-01 of *BS 7671* (that is, by colour or, in the case of certain multicore cables having numbered cores, by numbers).

Where colours are used, the phase conductor of any single-phase ac circuit should always be identified by red, regardless of the phase of the supply to which the circuit is connected. The only exception to this rule is that as an alternative to red, yellow and blue may be used in a large installation on the supply side of the final distribution boards. In three-phase ac circuits, the phase conductors should be identified by red, yellow or blue, according to the phase of supply to which they are connected. For dc circuits, refer to the colour requirements in Table 51A of *BS 7671*.

The use of the colour green is not permitted in new installation work, except in the colour combination green-and-yellow which is reserved for protective conductors.

In accordance with *BS 5467* and *BS 6724*, where cables having thermosetting insulation and numbered cores are used, the numbers 1, 2, and 3 should signify phase conductors, and the number 0, the neutral conductor. The same use of the numbers 1, 2, 3 and 0 applies to the numbered cores of paper-insulated cables to *BS 6480*. A core having the number 4 in this type of cable should be used as the fifth ('special-purpose') core, if required. Numbered cores are also provided in armoured PVC-insulated auxiliary cables which comply with *BS 6346* and which have five or more cores, but neither *BS 6346* nor *BS 7671* specify particular functions for the numbers.

Flexible cables and flexible cords:

Every core of a flexible cable or flexible cord should be identifiable by colour throughout its length in accordance with Regulations 514-07-01 and 514-07-02, and Table 51B of *BS 7671*. This means that cores used as phase conductors should generally be coloured brown (although certain other colours are permitted by Table 51B depending on the number of cores in the cable). A core used as neutral should be coloured blue, and a core used as a protective conductor should be coloured green-and-yellow.

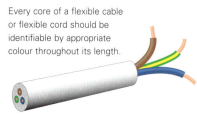

Every core of a flexible cable or flexible cord should be identifiable by appropriate colour throughout its length.

In a circuit which does not incorporate a neutral conductor, it is permitted to use a blue core for purposes other than neutral provided that it is not used as a protective conductor. If the blue core is used for another function, the coding 'L1, L2, L3', or other suitable marking where appropriate, should be used.

Where an indication of phase rotation is required, or it is required to distinguish the function of more than one conductor of the same colour, this should be achieved through the application of a numbered or lettered (not coloured-coded) sleeve to the core, using the coding 'L1, L2, L3' or other suitable coding where appropriate.

It is not permitted to use a flexible cable or flexible cord containing a core of the single colour green, the single colour yellow, or any bi-colour other than the colour combination green-and-yellow.

Bare conductors:

Reference should be made to Regulations 514-06-03 and 514-06-04 for the identification requirements for bare conductors

CABLES AND CONDUCTORS

Routing of cables in prescribed zones or within mechanical protection

Cables concealed or being concealed in the building fabric at a depth of 50 mm or less from the surface(s) of a wall or partition must be checked at the appropriate stage to verify that they are routed within prescribed zones as indicated below. *BS 7671* requires all cables that are not contained in these zones to incorporate an earthed metallic covering complying with the requirements for a cpc for the circuit concerned. Alternatively, cables are required to be protected by enclosure in earthed metal conduit, trunking or ducting, or by equivalent mechanical protection sufficient to prevent penetration by nails, screws and the like.

Zoning within a wall or partition:

A cable concealed within a wall or partition at a depth of less than 50 mm from the surface, unless otherwise provided with mechanical protection as previously described, must be positioned within the prescribed zones as shown. Both sides of the wall or partition must be considered. If the cable is less than 50 mm from either surface, it must be positioned within the prescribed zones as viewed from that side of the wall (where there are no prescribed zones available, cables must be otherwise protected as previously described), even if the cable does not connect to a point, accessory or item of switchgear on that side.

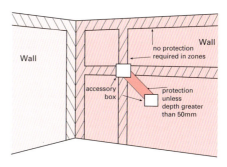

Cables which are run from points, accessories and switchgear must run in straight lines either vertically or horizontally.

Zoning within a floor or ceiling:

BS 7671 requires that where a cable is installed under a floor or above a ceiling, the cable must be run in such a position that it is not liable to be damaged by the floor or ceiling, or their fixings. Where the cable passes through a timber joist or batten etc, it must be at least 50 mm from the top, or bottom as appropriate (measured vertically), of the timber member (that is, the surface(s) of the joist or batten or other timber member into which fixings are liable to be made).See diagram on following page.

Connection of conductors

Every connection between conductors and between a conductor and equipment must be sound, both mechanically and electrically. The conductor(s) and insulation must be undamaged and the standard of workmanship must be such that the connection is likely to remain secure for the lifetime of the installation. The method of connection should take account of external influences such as moisture, vibration and thermal cycling, and there should be no appreciable mechanical strain on the terminations.

Connections must remain accessible during the lifetime of the installation. The only types of connection or joint which may be excused at a periodic inspection (but not during initial inspection) are compound-filled or encapsulated joints, brazed, welded or

Cables in floor joists

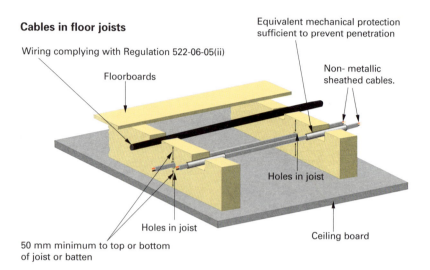

Wiring complying with Regulation 522-06-05(ii)

Equivalent mechanical protection sufficient to prevent penetration

Floorboards

Non- metallic sheathed cables.

Holes in joist

Holes in joist

Ceiling board

50 mm minimum to top or bottom of joist or batten

Example of incorrect use of ferrous enclosures

Warning!

This arrangement is
NOT
permitted

To avoid adverse ferromagnetic effects, conductors of an ac circuit should not be installed in separate steel conduits (see Regulation 521-02-01).

compression joints, and cold tail connections of underfloor and ceiling heating systems.

For purposes of protection against fire and thermal effects, every joint or connection in a live conductor or PEN conductor must be made in an enclosure complying with the requirements of Regulation 526-03 of *BS 7671*. It should be noted that this requirement applies not only to connections in low voltage circuits, but also to connections in extra-low voltage circuits. The enclosure must also provide adequate protection against mechanical damage and any other external influence or stress to which it may be subjected. The cores of sheathed cables from which the sheath has been removed must not extend outside the enclosure. In other words, the sheath, which is intended to provide mechanical protection, must extend into the enclosure.

Connections in protective conductors should not be overlooked in the inspection process, particularly those involving metallic wiring systems such as steel conduit and trunking or the metallic sheaths or armouring of cables. For example, a metal cable gland, coupled by a locknut to a hole in the sheet metal enclosure of a distribution board which has an insulating coating (such as paint), should not itself be regarded as providing a satisfactory electrical connection. In such a case, a sound connection should be provided by means of a conductor of copper or other suitable material of adequate cross-sectional area, connecting the gland earth tag washer to the earthing terminal or bar in the distribution board.

Erection methods

Adequate inspection of erection methods can usually be achieved only if the installation is inspected at appropriate intervals during its construction. Compliance with the relevant requirements of Chapter 52 of *BS 7671* must be considered at each stage. In general, the inspection should seek to confirm that all installed equipment and materials comply with the appropriate British Standard or equivalent, are properly selected and erected, and are not visibly damaged or defective so as to impair safety.

Matters to be checked should include, for example, the existence of manufacturer's markings or certification signifying compliance to British or Harmonised Standards; that cables and conduits etc are adequately supported at appropriate intervals and are not subject to mechanical stresses; that points, accessories and switchgear are securely fixed, and that items of equipment are not cracked, broken or otherwise defective.

Prior to the installation of cables, it should be confirmed that conduits are free from burrs, and that protection is provided at any sharp edges in wiring containment systems to prevent mechanical damage to the cables. Conduit bushes should be checked for tightness. The radii of bends in wiring containment systems should be checked to confirm that the cables will not be damaged during drawing in. It should be verified that the radii of bends in cables are not less than the minimum recommended for the particular type and overall diameter of cable.

Cable supports should be sufficient in number to prevent the cables from being damaged by their own weight or by electromechanical forces associated with fault conditions. There should be no appreciable strain on the terminals.

Single cables of ac circuits run in steel conduit or any other form of steel enclosure should be installed so that all phase, neutral and, where provided, separate protective conductors are together within the same enclosure. This requirement also applies to cables passing through holes in ferromagnetic enclosures (such as steel trunking). Single-core cables armoured with steel wire or tape are not permitted in ac circuits due to the adverse thermal and electromagnetic effects associated with such use.

Cables which are exposed to direct sunlight or to the effects of ultra-violet radiation should have been selected for such application. Generally, PVC-sheathed cables may require some protection from ultra-violet radiation, but bare mineral insulated cables usually do not. For all types of cable, advice should be sought from the designer of the electrical installation, if in doubt.

Selection of conductors for current-carrying capacity and voltage drop

The sizes and types of all the installed cables, and the conductors of any bus-bar trunking system and the like, should be checked against the installation design information to confirm that they are as intended by the designer. It is a requirement of *BS 7671* that this design information is made available to those carrying out the inspection and testing of new installation work, irrespective of whether the design was carried out by the contractor or by another party such as a consulting engineer.

Presence of fire barriers, suitable seals and protection against thermal effects

Where the wiring system passes through an element of building construction (such as a wall or floor) having a specified fire resistance, the space around the wiring system must be sealed to a degree of fire resistance which is no less than that of the building element concerned.

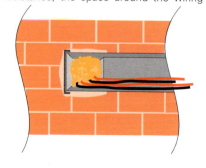

Similarly, where a wiring system having an internal space, such as trunking, passes through an element of a building having a specified fire resistance, the wiring system must be sealed internally to the appropriate degree of fire resistance. The exception to this requirement is that a non-flame-propagating wiring system, with an internal cross-sectional area of not more than 710 mm², need not be sealed internally. This is equivalent to the internal cross-sectional area of a 32 mm diameter conduit.

Internal barriers may be made from a type of intumescent packing which expands to fill the space when heated to the temperatures likely to occur under fire conditions. Solidifying foam may also be used.

For a new installation, the presence of the required fire barriers must be verified during the construction stage, when access to all the wiring system penetrations should be readily available.

Protection against thermal effects, such as the proximity of hot surfaces to electrical equipment, must be checked. Sufficient clearance or an appropriate form of permanent heat shield or barrier may need to be provided during construction to prevent adverse effects on wiring systems.

GENERAL ITEMS

Presence and correct location of appropriate devices for isolation and switching

The inspector should check the installation against the design information used for its construction, to confirm that all devices for isolation and switching are present and correctly located (and identified), as intended by the designer. It is a requirement of *BS 7671* that this information is made available to those carrying out the inspection and test (see Regulation 711-01-02).

'Isolation and switching' is the general term used to cover four basic functions, namely isolation, switching off for mechanical maintenance, emergency switching and functional switching. These functions are considered separately for the purposes of inspection, as described below. It should be remembered, however, that two or more of the functions may be performed by a single device, provided that the arrangement and characteristics of the device satisfy all the requirements of *BS 7671* for the particular functions concerned.

Whilst checking the items referred to below, it should also be confirmed that all devices for isolation and switching are suitable for the currents they will have to carry and, where appropriate make or break, under normal and fault conditions.

Isolating devices:

The purpose of isolation is to enable skilled persons to carry out work on, or in the vicinity of, parts that are normally live in service without risk of injury from electric shock from those parts. *BS 7671* requires that every circuit must be capable of being isolated.

It should be confirmed that the installation has a main linked switch or linked circuit-breaker positioned as near as practicable to the origin. This is intended as a means of switching the supply on load, and as a means of isolation. Where the installation is supplied from more than one source, a main switch or main circuit-breaker is required for each source.

The presence of the correct number of poles in the main switch or circuit-breaker should be verified.

The device is required to interrupt all live conductors of the supply except as follows:

- In a TN-S or TN-C-S system, the neutral conductor generally need not be switched where this can reliably be regarded as being at earth potential provided that, in all cases, a single-phase main switch intended for operation by unskilled persons (such as in a consumer unit) interrupts all live conductors.

- In a TN-C system, the PEN conductor must not be isolated or switched.

Where the main switch or circuit-breaker does not interrupt the neutral conductor, *BS 7671* requires provision to be made for the disconnection of the conductor for testing purposes.

Although the main switch or circuit-breaker provides a means of isolation for the whole installation, it is probable that other isolating devices need to be present so that individual items of equipment, circuits or groups of circuits can be isolated independently. The presence of these other devices, where appropriate, should be verified. For example, *BS 7671* requires that every motor circuit is provided with a disconnector (isolator) to disconnect the motor and all equipment, including any automatic circuit-breaker, associated with it. It is likely that, for operational reasons, additional isolating devices will form part of the design of any large installation, so that a section can be safely worked on without necessarily having to isolate the whole installation, which would probably be inconvenient to the users.

The additional isolating devices must interrupt all live conductors except, generally, the neutral conductor in TN-S and TN-C-S systems where this can reliably be regarded as being at earth potential. In a TN-C system, the PEN conductor must not be isolated or switched.

It should be checked that each device used for isolation is of a type which *BS 7671* permits for this purpose. Such devices include an isolator (disconnector), isolating switch (switch disconnector), suitable circuit-breaker, plug and socket-outlet, (withdrawable) fuse, or disconnectable link.

Devices for isolation are required to have a contact separation distance not less than that determined for an isolator conforming to *BS EN 60947-3*. Unfortunately, this standard is not simple to apply. The position of the contacts must be either externally visible or clearly and reliably indicated when the specified isolating distance has been attained in each pole. As a rule of thumb, a contact separation of 3 mm is generally considered to be sufficient, but if there is any doubt as to whether a device is designed to provide isolation, its suitability must be confirmed by enquiry.

All isolating devices should have been selected and/or installed in such a way as to prevent unintentional reclosure (such as by mechanical shock or vibration). Means should have been provided to secure, in the open position, any isolating devices which are placed remotely from the equipment to be isolated, and to secure any off-load isolating devices, to prevent inadvertent or unauthorised operation.

Devices for switching off for mechanical maintenance:

The purpose of switching off for mechanical maintenance is to enable persons to safely replace, refurbish or clean lamps, and to maintain non-electrical parts of electrical equipment, plant and machinery.

It should be verified that a means of switching off for mechanical maintenance is provided where any of the above operations may involve a risk of burns, or a risk of injury from mechanical movement. For example, a means of switching off for mechanical maintenance would be required to enable maintenance personnel to clean luminaires having lamps that may cause burns, or to enable a mechanical fitter to work on the normally moving parts of an electrically-driven machine. It should be noted, however, that a suitable means of isolation would also be required if any of these operations were likely to involve a need to work on or near live parts, such as could be the case if the maintenance operations involved dismantling the equipment.

The devices used for switching off for mechanical maintenance should be checked to ensure that they are of the types permitted for this function by *BS 7671*, namely switches or suitable socket-outlets. Switches must be capable of cutting off the full load current of the relevant part of the installation, and either have an externally-visible contact gap, or give a clear and reliable indication of the OFF or OPEN position when each pole of the contacts is properly opened. Plugs and socket-outlets are permitted for such purposes only if their rating does not exceed 16 A.

Devices for switching off for mechanical maintenance must be suitably positioned for ready operation. This often means that the device has to be positioned adjacent to the equipment requiring maintenance. Unless the device is to be continuously under the control of the person performing the maintenance, suitable provisions (such as locking off facilities on the device) must be available to prevent the equipment from being reactivated during mechanical maintenance. In any event, each device must have been selected and/or installed such that unintentional reclosure by mechanical shock and so on is unlikely to occur.

It should be confirmed that each device for switching off for mechanical maintenance is connected in the main supply circuit. Such a device may only be connected in the control circuit where supplementary precautions have been taken to provide a degree of safety equivalent to that of interruption of the main supply.

Devices for emergency switching:

The purpose of emergency switching is to remove, as quickly as possible, danger which may have occurred unexpectedly.

It should be verified that a means of emergency switching is provided for every part of the installation for which it may be necessary to cut off the supply rapidly to prevent or remove danger. For example, an emergency switching device would be required for a drilling machine in a factory machine shop. Where an electrically-driven machine may give rise to danger, *BS 7671* requires that a means of emergency stopping be provided (that is, emergency switching intended to stop the operation whilst, if necessary for safety, leaving certain parts of the equipment energised).

Emergency switching devices must be suitably positioned for ready operation and must not be sited where access to them is likely to be impeded in any foreseen emergency. The operating means (such as a handle or push-button) must be clearly identifiable; must be capable of being restrained in the OFF or STOP position; and should preferably be coloured red. *BS 7671* forbids the selection of a plug and socket-outlet as an emergency switching device.

It should be confirmed that the means of interrupting the supply is capable of cutting off the full load current of the relevant part of the installation. The means may, for example, be a contactor used in conjunction with the emergency switching device. Due account must be taken of any stalled motor conditions.

The inspector should confirm that the means of emergency switching acts as directly as possible on the appropriate supply conductors as required by *BS 7671*. There should be no unnecessary relays etc in the emergency switching arrangement. Where a risk of electric shock is involved, the means should interrupt all live conductors except that the neutral of a TN-S or TN-C-S system need not be interrupted, where this is reliably regarded as being at earth potential.

Devices for functional switching:

Devices for functional switching are dealt with here for the sake of convenience, but they are actually the subject of the final item in the Schedule of Items Inspected in the Electrical Installation Certificate, Domestic Electrical Installation Certificate or Periodic

Inspection Report. It is against that item that the corresponding tick should be placed, once the provisions have been verified.

The inspector should verify that the devices to switch 'on' and 'off' or to vary the supply of electrical energy for normal operating purposes are appropriate in quantity, type, function and location.

Functional switching is required for any circuit or part of a circuit that may require to be controlled independently of other parts of the installation. Devices for functional switching may be single-pole or multi-pole; they need not control all live conductors but must not solely control the neutral. Suitable devices include switches, plugs and socket-outlets rated at up to 16 A (except for control of dc) and semiconductor devices. Off-load isolators, fuses or links etc must not be used for functional switching.

Adequacy of access to switchgear and other equipment

It must be confirmed that adequate and safe means of access and working space have been afforded to every piece of electrical equipment requiring operation, inspection, testing or maintenance.

Open type switchboards and other types of equipment which have exposed live parts are not commonly encountered, but access to such equipment should be restricted to skilled or instructed persons. Regulation 471-13-02 requires that the dimensions of each passageway and working platform for such equipment must be adequate to allow persons, without hazard, to operate and maintain the equipment, pass one another as necessary with ease, and to back away from the equipment.

Appendix 3 of the *Memorandum of Guidance on the Electricity at Work Regulations* advises that, for low voltage applications, the working space for the open type of equipment should have a height of not less than 2.1 m and a clear width measured from the bare conductor of not less than 0.9 m.

Particular protective measures for special installations and locations

The special installations and locations referred to in this section are those where the risks associated with electric shock are increased due to wetness; absence of, or minimal clothing; presence of earthed metal; arduous conditions; and where other risks, such as that of fire, may also be increased. Some of these locations and installations are included in Part 6 of *BS 7671*, where the particular protective measures associated with them are also given. These measures modify or add to the general requirements for safety.

Inspection, Testing, Certification and Reporting

PIR/ XXXXXXX

Original (to the person ordering the work)

SCHEDULE OF ITEMS INSPECTED (See Section 712 of BS 7671: 1992) † See note below

Methods of protection against electric shock

a. Protection against both direct and indirect contact:

    (i) SELV

    (ii) Limitation of discharge of energy

b. Protection against direct contact:

    (i) Insulation of live parts

    (ii) Barriers or enclosures

    (iii) Obstacles

    (iv) Placing out of reach

    (v) PELV

    (vi) Presence of RCD for supplementary protection

c. Protection against indirect contact:

    (i) EEBAD including:

    Presence of earthing conductor

    Presence of circuit protective conductors

    Presence of main equipotential bonding conductors

    Presence of supplementary equipotential bonding conductors

    Presence of earthing arrangements for combined protective and functional purposes

    Presence of adequate arrangements for alternative source(s), where applicable

    Presence of residual current device(s)

    (ii) Use of Class II equipment or equivalent insulation

    (iii) Non-conducting location:
    Absence of protective conductors

    (iv) Earth-free equipotential bonding:
    Presence of earth-free equipotential bonding conductors

    (v) Electrical separation

Prevention of mutual detrimental influence

a. Proximity of non-electrical services and other influences

b. Segregation of Band I and Band II circuits or Band II insulation used

c. Segregation of safety circuits

Identification

Presence of diagrams, instructions, circuit charts and similar information

Presence of danger notices and other warning notices

Labelling of protective devices, switches and terminals

Identification of conductors

Cables and Conductors

General

Presence and correct location of appropriate devices for isolation and switching

Adequacy of access to switchgear and other equipment

Particular protective measures for special installations and locations

Connection of single-pole devices for protection or switching in phase conductors only

Correct connection of accessories and equipment

Presence of undervoltage protective devices

Choice and setting of protective and monitoring devices (for protection against indirect contact and/or overcurrent)

Selection of equipment and protective measures appropriate to external influences

Selection of appropriate functional switching devices

Choice and setting of protective and monitoring devices (for protection against indirect contact and/or overcurrent)

Selection of equipment and protective measures appropriate to external influences

Selection of appropriate functional switching devices

SCHEDULE OF ITEMS TESTED (See Section 713 of BS 7671: 1992) † See note below

External earth fault loop impedance, Z_e

Installation earth electrode resistance, R_A

Continuity of protective conductors

Continuity of ring final circuit conductors

Insulation resistance between live conductors

Insulation resistance between live conductors and earth

Site applied insulation

Protection by separation of circuits

Protection against direct contact by barrier or enclosure provided during erection

Insulation of non-conducting floors or walls

Polarity

Earth fault loop impedance, Z_s

Operation of residual current devices

Functional testing of assemblies

† **All data-entry boxes must be completed.** To provide a positive indication that an inspection or a test has been carried out, insert either a 'Yes' or a '✓'. Where an inspection or a test is not relevant to the installation, insert 'N/A' meaning 'Not Applicable'. Exceptionally, where a limitation on a particular inspection or test has been agreed, and has been recorded in Section D, the appropriate data-entry box(es) must be completed by inserting 'LIM', indicating that an agreed limitation has prevented the inspection or test being carried out.

Page 4 of ☐

This form is based on the model shown in Appendix 6 of BS 7671: 1992, as amended 1997
Published by the National Inspection Council for Electrical Installation Contracting © Copyright NICEIC (Jan 2000)

The installations and locations included in Part 6 of *BS 7671* - such as locations containing a bath tub or shower basin, and locations where equipment has high earth leakage currents - are not the only ones where there may be increased risk of electric shock. The designer of an installation may have identified other locations of increased shock risk and decided to use additional measures such as 30 mA RCD protection for certain equipment, supplementary bonding, or a reduction of maximum fault clearance time, in accordance with Regulation 471-08-01.

The inspector must carefully examine all of the special locations and installations within the particular installation being inspected, including any additional ones identified by the designer, to verify that the associated protective measures are present in accordance with the relevant Sections in Part 6 of *BS 7671* and the design information for the installation.

'Special installations and locations' are fairly common, and the inspector will need to think carefully before putting an N/A in the relevant box on the Electrical Installation Certificate, Domestic Electrical Installation Certificate or Periodic Inspection Report. It should be remembered that the ordinary bath or shower room is a special location and a tick in this box is normally applicable for every domestic installation.

Connection of single-pole devices for protection or switching in phase conductors only

This is a visual check to verify that all single-pole switching and control devices are connected in the phase conductor only. No single-pole switching device, circuit-breaker, fuse, solid link which can be removed without the use of a tool or key, or the like should interrupt or control the neutral conductor. This check should be made in addition, and not as an alternative, to the polarity test which will be conducted later in the verification process.

Correct connection of accessories and equipment

The inspector should examine the installation to verify that the conductors at accessories and other items of equipment are connected to the correct terminals, and that the connections are well made both electrically and mechanically.

Presence of undervoltage protective devices

The inspector should examine the installation for undervoltage protective devices which are provided for the purpose to disconnect equipment in the event of a significant reduction in the supply voltage. These devices are not uncommon. They are found, for instance, in starter units controlling and protecting electric motors. They may also be found in emergency stop systems, and in automatic control systems to bring stand-by supplies into operation.

Choice and setting of protective and monitoring devices
Examples of devices which should be checked:-

The BS (EN) number, type (eg general-purpose or selective) and rated residual current of each RCD.

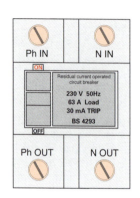

The BS (EN) number, type and nominal current of each fuse.

The BS (EN) number, type and nominal current of each MCB.

The BS (EN) number, type (characteristics) and, where appropriate, settings of each moulded-case circuit breaker.

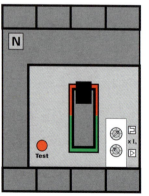

Note: Drawings not to scale.

The inspector should check that any undervoltage devices installed conform to the design requirements for the installation and Chapter 45 of *BS 7671.*

The presence of any necessary arrangements to prevent automatic reclosure of undervoltage devices should also be verified. Such arrangements are required where the reclosure of an undervoltage device is likely to cause danger (for example, where the unexpected restarting of a machine could cause danger).

Choice and setting of protective and monitoring devices (for protection against indirect contact and/or overcurrent)

This item is concerned with fuses, circuit-breakers, residual current devices and any current monitoring relays which control separate circuit-breakers. Their current ratings, settings (where applicable) and characteristics must be checked to verify conformity with the design information for the installation.

For each fuse and circuit-breaker, a check should be made of not only the nominal current but also the type (such as gG for a fuse to Parts 2 and 6 of *BS 88*, or Type B for a circuit-breaker to *BS EN 60898*).

For each moulded case circuit-breaker, a check should be made of the nominal current or current setting, where appropriate, and the time/current characteristics. It should be noted that on some moulded case circuit-breakers, both the time/current characteristics and the current setting are adjustable. These must be checked against the design information.

For each residual current device, a check should be made of the rated residual operating current. The time/residual current characteristics should also be checked where the device is to be used for supplementary protection against direct contact, or where an intended time delay is incorporated.

The current and time settings of each current monitoring relay should be checked against the design.

Selection of equipment and protective measures appropriate to external influences

The installation should be inspected during construction and on completion to confirm that the installed equipment and protective measures are suitably selected and erected to operate safely, given whatever external influences are reasonably likely to occur during the lifetime of the installation.

An external influence is defined in Part 2 of *BS 7671* as any influence external to an electrical installation which affects the design and safe operation of that installation'. Examples of external influences include ambient temperature, external heat sources, solar radiation, impact, vibration, the presence of water, high humidity, solid foreign bodies, corrosive or polluting substances, or other factors such as structural movement.

Where there is doubt about the types or severity of the external influence(s) which are likely to act on an installation, reference should be made to the designer of the installation, who should have made an assessment of these influences as part of the design process in accordance with Part 3 of *BS 7671*.

BS 7671 requires that all wiring and other equipment must be of a design appropriate to the situation in which it is to be used, and/or that its mode of installation must take account of the conditions likely to be encountered. Moreover, any equipment which does not meet these requirements by its own construction must be provided with appropriate additional protection. For example, in an installation in industrial premises where there are dusty or wet conditions, the selected luminaires and wiring system may, by their own construction, be suitable for the arduous conditions, whereas the distribution boards may be of a standard pattern, but housed within suitable cupboards or additional electrical enclosures appropriate for both the equipment and the conditions. It should be checked that and additional protection against external influences, for example by an enclosure, does not cause overheating of the protected equipment, impair its operation, nor impede its inspection, testing and maintenance.

It is important to recognise that the risk of mechanical damage and other damaging influences such as dust and water may be greater during the construction process than when the installation is in service. Care must therefore be taken to ensure, so far as is reasonably practicable, that any damaged cables and the like are discovered and replaced before they are concealed in the fabric of the building.

Selection of appropriate functional switching devices

This item was covered earlier in this Chapter, at the end of the section relating to 'presence and location of appropriate devices for isolation and switching'.

Chapter

8

Testing

Sequence of Tests	
For initial testing	**For periodic testing**
(i) Before the supply is connected, or with the supply disconnected as appropriate • Continuity of protective conductors, main and supplementary bonding. • Continuity of ring final circuit conductors. • Insulation resistance. • Site-applied insulation. • Protection by separation of circuits. • Protection by barriers or enclosures provided during erection. • Insulation of non-conducting floors and walls. • Polarity. • Earth electrode resistance. **(ii) With the electrical supply connected (re-check polarity before further testing)** • Prospective fault current. • Earth fault loop impedance. • Residual current operated devices. • Functional test of switchgear and controlgear.	(i) The following are generally applicable • Continuity of all protective conductors (including earthed equipotential bonding conductors). • Polarity. • Earth fault loop impedance. • Insulation resistance. • Operation of devices for isolation and switching. • Operation of residual current devices. • Prospective fault current. **(ii) Where appropriate, the following tests must also be undertaken** • Continuity of ring circuit conductors. • Earth electrode resistance. • Manual operation of overcurrent protective devices other than fuses. • Electrical separation of circuits - insulation resistance. • Protection by non-conducting floors and walls - insulation resistance.

8 TESTING

SAFETY

Electrical testing inherently involves a degree of risk. Persons carrying out electrical testing have a duty to ensure the safety of themselves and others from any possible danger resulting from the tests or from access to live parts. This requires strict adherence to safe working practice, including checking that the test equipment and leads are safe and suitable for the intended purpose, that the equipment to be tested is safe for the tests, and that the working environment does not impair safety. It is important to appreciate there is a potential danger of electric shock not only from electricity supplies, but also from voltages generated within test instruments, from stored energy devices such as uninterruptible power supplies, and from capacitive loads. Particular care is needed where there are alternative sources, such as from standby generators and uninterruptible power supplies.

General advice on such safety matters is given in Section 1.2 of *IEE Guidance Note 3: Inspection and Testing*, and in *HSE Guidance Note GS 38: Electrical Test Equipment for use by Electricians*.

SEQUENCE OF TESTS

As mentioned in Chapter 5, *IEE Guidance Note 3* gives a different sequence for periodic testing to that which is given in *BS 7671* for new installation work (including work involving an addition or an alteration). The comparison table given in Chapter 5, listing both sequences, is repeated opposite for ease of reference.

NOTE: The tests, as applicable to the particular installation, should be conducted in the appropriate sequence shown opposite for reasons of safety. Except for the measurement of maximum prospective fault current, the order in which the tests are covered in this Chapter is the order in which they appear in the 'Schedule of Items Tested' in the NICEIC Electrical Installation Certificate and NICEIC Periodic Inspection Report and, with certain omissions, the Domestic Electrical Installation Certificate. This is not necessarily the same as the order in which the tests should be conducted.

If any test carried out during the initial verification of installation work indicates a failure to comply, that test, and any preceding test which may have been affected by the defect, must be repeated after the defect has been rectified.

If any test carried out during periodic inspection and testing of an installation indicates a failure to comply, then (unless there is evidence of real and immediate danger which requires urgent action as discussed in Chapter 5) the result of that test must be duly recorded on the Periodic Inspection Report together with appropriate observations, recommendations and, where appropriate, comments in the summary section of the report.

Measurement of prospective fault current using a two-lead test instrument

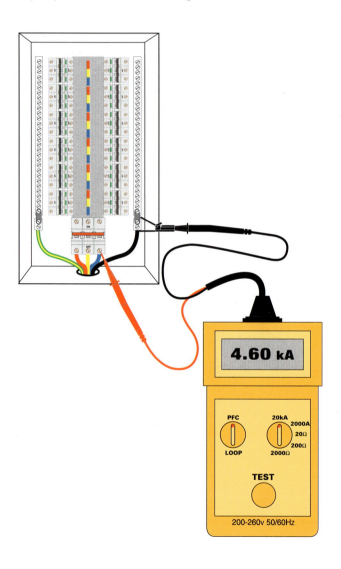

MEASUREMENT OF MAXIMUM PROSPECTIVE FAULT CURRENT

The maximum prospective fault current is one of the 'supply characteristics and earthing arrangements' to be recorded on the second page of the Electrical Installation Certificate, first page of the Domestic Electrical Installation Certificate, or the third page of the Periodic Inspection Report.

This is the current that would flow in the event of a fault at that location, either between live conductors or between live conductors and protective earth, whichever is the greater. The value can only be measured by testing at the origin of the installation, which is at the position where the electrical energy is delivered. The tests should, therefore, be conducted at the main switch or at other switchgear connected directly to the tails from the Public Electricity Supplier's metering equipment (or at the main low voltage switchgear connected to the tails from the consumer's sub-station transformer if the supply is delivered at high voltage).

Take particular care during the testing process, as fault conditions are most severe at the origin of an installation, where this test is being performed.

It would be incorrect to make these measurements at a position remote from the origin. For example, if the measurements were made at an item of switchgear fed by a distribution circuit, the measured value of prospective earth fault current at that point would not be the maximum for the installation.

The earthing conductor, main bonding conductors and circuit protective conductors should all be in place during this test, because the presence of these and any other parallel paths to earth may reduce the impedance of the fault loop and so increase the prospective fault current.

The instrument to be used is an earth fault loop impedance test instrument, with or without a prospective fault current range. A set of test leads is required, having two or three probes to suit the requirements of the particular test instrument.

The figure opposite shows a two-lead test instrument connected across the phase and neutral of an incoming supply. Note that for measuring a phase to earth value, the instrument must be connected across the phase and main earthing terminal.

Measurement of prospective fault current using a three-lead test instrument

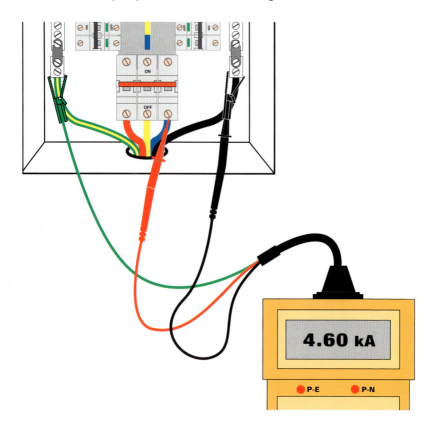

The figure opposite shows a three-lead test instrument connected across the phase, neutral and earth of an incoming supply. Note that the connections may need to be different for some tests - refer to the instrument maker's instructions.

Procedure:

- Check that the test instrument, leads, probes and crocodile clips (if any) are suitable for the purpose, and in good serviceable condition.

- Select the appropriate range and scale (for example, prospective fault current 20 kA).

- Observing all precautions for safety, connect the instrument to the incoming energised supply to measure a phase to neutral value.

- Check the polarity indicator (if any) on the instrument for correct connection.

- Press the button.

- Record the reading.

- Repeat the above steps, as appropriate, with the instrument connected for measurement of the phase to earth value.

If the test instrument does not have a prospective fault current range, the readings given by the above procedure are fault loop impedances (in ohms). To convert each of these readings into a prospective fault current, divide them into the measured value of phase to neutral voltage.

For example, if the voltage measured at the time of the test is 230 V, and the measured value of fault loop impedance between phase and neutral at the origin is 0.05 Ω:

Prospective fault current (phase to neutral) = 230/0.05 = 4600 A (or 4.6 kA)

(This value would have been given directly if the instrument had a prospective fault current range).

The same procedure is now repeated, taking readings between phase and earth to determine the prospective phase to earth fault current.

The higher of the two prospective fault currents (phase to neutral or phase to earth) should be recorded on the certificate or report.

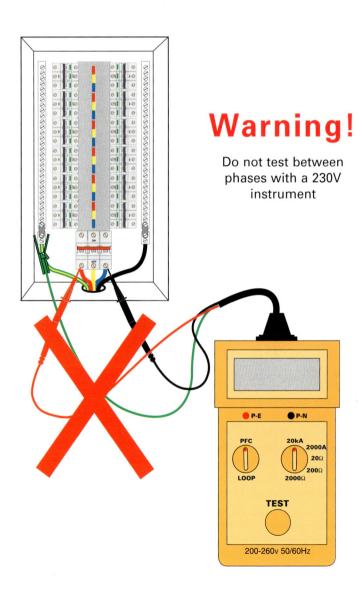

Warning!

Do not test between
phases with a 230V
instrument

Three-phase supplies

Unless a test instrument designed to operate at 400 V is available, it will be necessary to calculate the prospective fault current between phases.

Where an installation is supplied by two or more phases, the maximum prospective fault current is likely to be between phase conductors. If this is found to be the case, then this is the value to be recorded on the certificate or report.

This can be done by using the measured value of the phase to neutral prospective fault current (obtained as described on the preceding pages). There is a complex relationship between the phase to neutral and the phase to phase values of prospective fault current, which is beyond the scope of this book. As a first approximation, which tends to err on the safe side, the prospective fault current phase to phase can be taken as twice that of the prospective fault current phase to neutral.

Using the previous example, phase to neutral prospective fault current multiplied by two gives:

$$4.6 \text{ kA} \times 2 = 9.2 \text{ kA}$$

(for the calculated value of prospective fault current between phases).

Why do we have to determine prospective fault current?

A fault current protective device, particularly when fitted close to the supply intake position, must generally be capable of successfully interrupting the current which would flow under conditions of short-circuit or earth fault. Otherwise there may be a serious risk of arcing, damage and fire, particularly for circuits having high prospective fault currents. Accordingly, the short-circuit capacity of a device must be no less than the prospective fault current at the point in the circuit at which it is installed. The only exception to this rule is where a device having a lower short-circuit capacity is installed in conjunction with what is sometimes called 'back-up' protection (for example, the design might require a circuit-breaker to be backed up by an appropriate HBC fuse).

Typical circuit-breaker identification

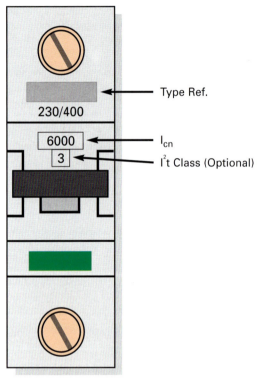

Front view only

The short-circuit capacity of typical devices are as follows:

Circuit-breakers to *BS EN 60898* have a value in a small rectangular box indelibly marked on the device, for example:

$$\boxed{6000}$$

This is the 'rated short-circuit capacity' of the device, I_{cn}, assigned by the manufacturer (in amperes, not kA). The value is determined on the basis of type-tests. It is the ultimate short-circuit capacity, and the device may be designed to break a fault of this magnitude only once.

The maximum fault current the device is designed to break more than once, called the 'service short-circuit capacity', I_{cs}, may be less than the ultimate short-circuit capacity marked on the device, and the manufacturer's data should be consulted for details.

A circuit-breaker to *BS 3871* will have an 'M' number marked on it. This indicates its short-circuit capacity in kA.

For example: M4 denotes 4000 A (4 kA), M9 denotes 9000 A (9 kA).

Fuses are not so straightforward:

Typical values are:

> *BS 3036* - 1 kA to 4 kA (depending on duty rating)
>
> *BS 1361* - 16.5 kA
>
> *BS 88* - 40 kA to 80 kA (depending on duty)

The above values are given only as a guide, and the manufacturer's data or other relevant information for the particular device should be consulted.

NOTE: Overcurrent protective devices having current ratings up to 50 A incorporated in a consumer unit conforming to Part 3 of *BS EN 60439: 1991* (Annex ZA of Corrigendum April 1994) are considered adequate for prospective fault current, provided the consumer unit is fed by a service cut-out having an HBC fuse to *BS 1361* Type II, rated at not more than 100 A, on a single-phase supply of nominal voltage up to 250 V.

Inspection, Testing, Certification and Reporting

Measurement of earth fault loop impedance

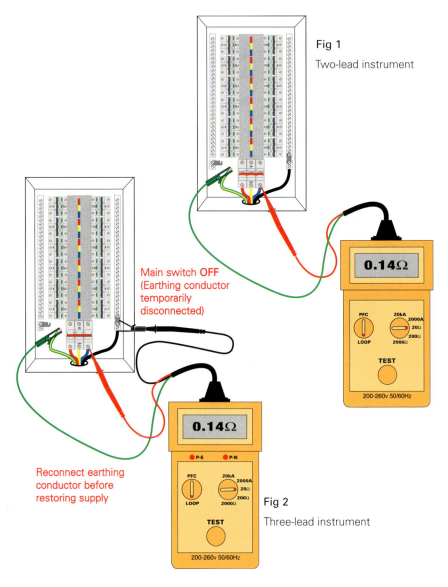

Fig 1

Two-lead instrument

Main switch **OFF**
(Earthing conductor
temporarily
disconnected)

0.14Ω

PFC   20kA   2000A
       20Ω
LOOP   2000Ω   200Ω

TEST

200-260v 50/60Hz

Reconnect earthing
conductor before
restoring supply

0.14Ω

● P-E   ● P-N

PFC   20kA   2000A
       20Ω
LOOP   2000Ω   200Ω

TEST

200-260v 50/60Hz

Fig 2

Three-lead instrument

MEASUREMENT OF EXTERNAL EARTH FAULT LOOP IMPEDANCE, Z_e

The external earth fault loop impedance, Z_e, is another of the 'supply characteristics and earthing arrangements' to be recorded on the second page of the Electrical Installation Certificate, on the first page of the Domestic Electrical Installation Certificate or the third page of the Periodic Inspection Report.

Z_e is the earth fault loop impedance of that part of the system which is external to the installation. Like the maximum prospective fault current, Z_e can only be measured by testing at the origin of the installation.

Furthermore, before the test, the earthing conductor must be disconnected from the main earthing terminal or otherwise separated from all parallel earth paths such as bonding and circuit protective conductors, as the purpose of the test is to prove that the intended means of earthing is present and of an acceptable ohmic value. Any parallel earth paths present during the test would result in a false reading of Z_e, and perhaps conceal the fact that the intended means of earthing is defective, or even non-existent.

For safety reasons, it is essential that the entire installation is isolated from the supply before the earthing conductor (or, if necessary, any other protective conductors) are disconnected from the main earthing terminal, and that the installation remains isolated for the duration of the testing.

On completion of testing and before the installation is re-energised, the earthing conductor or other protective conductors disconnected to facilitate the measurement of Z_e must be reconnected.

Take particular care during the testing process, as the fault conditions are most severe at the origin of the installation, where this test is being performed.

The instrument to be used is an earth fault loop impedance test instrument complete with a set of test leads, having two or three probes to suit the particular instrument. In the **two-lead** version, connection is made between phase and earth of the incoming supply (see Fig 1). In the **three-lead** version it is necessary to make a connection to the neutral of the incoming supply as well as to phase and earth (see Fig 2).

If the supply is derived from a three-phase system and there is no neutral present, then the neutral lead of a three-lead test instrument should be connected to the earthing conductor.

Some instruments have a facility for polarity indication. Before proceeding with the test, it is necessary to follow the manufacturer's instructions with regard to correct polarity and test procedure. Do not proceed with the test unless the indications on the instrument show that it is safe to do so.

Where the supply is provided with an earth connection from the Public Electricity Supply network, such as from a TN-S or TN-C-S system, it is unlikely that Z_e will exceed one ohm, and the lowest range of the instrument will be the most appropriate (usually 0 to 20 Ω).

Inspection, Testing, Certification and Reporting

Typical maximum values published by Public Electricity Suppliers are:

TN-S   0.80 Ω

TN-C-S   0.35 Ω

In a TT system, where the means of earthing for the installation is provided by an installation earth electrode, the value of Z_e is likely to fall within a higher range of the test instrument.

Procedure:

- Open the main switch to disconnect the whole of the installation from the source of supply, and secure the switch in the open position.

- Disconnect the earthing conductor from the main earthing terminal or otherwise separate it from all main bonding conductors and circuit protective conductors.

- Check that the test instrument, leads, probes and crocodile clips (if any) are suitable for the purpose and in good serviceable condition.

- Observing all precautions for safety, apply the test probes.

- Check the polarity indicator for correct connection.

- Press the test button.

- Record the result.

- **Reconnect the earthing conductor, main bonding conductors and circuit protective conductors immediately after the tests have been completed and before the supply is restored to the installation.**

Why do we have to measure external earth fault loop impedance?

In order to verify compliance with *BS 7671*, knowledge of the earth fault loop impedance is required for each circuit that relies for indirect contact protection on Earthed Equipotential Bonding and Automatic Disconnection of supply (EEBAD).

Without a sufficiently low earth fault loop impedance, the magnitude of the earth fault current will not be sufficient to cause the protective device to automatically disconnect the circuit within the maximum time permitted to provide protection against electric shock (indirect contact) or to provide overcurrent protection against earth fault currents. The external earth fault loop impedance of an installation forms part of the earth fault loop impedance of every one of its circuits. Thus *BS 7671* requires that Z_e be determined. Importantly, establishing Z_e by measurement also verifies that the intended means of earthing is present.

INSTALLATION EARTH ELECTRODE RESISTANCE, R_A

A TT system will always be equipped with an installation earth electrode as its means of earthing. The function of the earth electrode is to make electrical contact with the general mass of earth to provide a suitable path for earth fault current to return to the source (such as via the earth electrode at the distribution transformer). Measurement of the resistance of the installation earth electrode to the general mass of earth provides information necessary for verifying that the earthing arrangement satisfies the requirements of both *BS 7671* and the particular installation design. The method of measurement and the measured resistance value need to be recorded, together with the type of electrode and its location, in the appropriate boxes under the heading of 'particulars of the installation at the origin' on the second page of the Electrical Installation Certificate, on the first page of the Domestic Electrical Installation Certificate or the third page of the Periodic Inspection Report.

Two methods of measuring the resistance of an earth electrode are recognised in *IEE Guidance Note 3*:

- Method 1: using a proprietary earth electrode test instrument.
- Method 2: using an earth fault loop impedance test instrument.

Where sufficient and suitable ground area is available, a proprietary earth electrode test instrument may be used. However, where there is insufficient space, or where hard surfaces make the use of such a method impracticable, the use of an earth fault loop impedance test instrument may be appropriate.

For safety reasons, these methods require the installation to be isolated from the supply before the means of earthing is disconnected.

Method 1

Having first isolated the installation, the earth electrode should preferably be disconnected from the earthing conductor which connects it to the main earthing terminal. If this is not possible, the earthing conductor must be disconnected from the main bonding conductors and other protective conductors at the main earthing terminal, to remove any parallel paths to the general mass of earth. If such parallel paths are not removed before the measurement is made, a false (low) reading of the resistance of the electrode to earth will be obtained.

If the test instrument is a four-terminal device, the terminals marked C1 and P1 are linked together and connected to the electrode under test. The remaining two terminals are connected to temporary electrodes as shown overleaf. Where the instrument is a three-terminal device, the corresponding configuration also shown overleaf. Many proprietary instruments have a facility for checking the resistance of

Measurement of installation earth electrode resistance

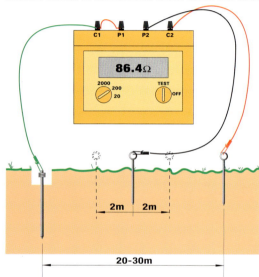

Method 1 (using a four-terminal proprietary earth electrode test instrument)

Method 1 (using a three-terminal proprietary earth electrode test instrument)

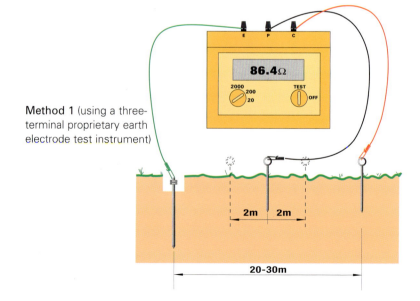

the temporary electrodes. Where the resistance of a temporary electrode is found to exceed the limit stated in the instrument manufacturer's instructions, the resistance should be reduced to a value within the stated limit.

A sufficient reduction in the resistance of the temporary electrodes may be achieved by driving longer temporary rods into the ground, or by watering the ground in the immediate vicinity of the temporary rods. The ground adjacent to the installation earth electrode must **not** be treated with water or brine, as this would make the resistance measurement invalid for the purpose of verification or reporting.

Where possible, the test instrument should be situated close to the installation earth electrode under test. If the test instrument has to be situated some considerable distance from that electrode, two separate leads should be run out from C1 and P1 to minimise the effect of lead resistance, with the link being formed at the installation earth electrode.

In principle, the instrument passes a current through the ground between the two current terminals C1 and C2. At the same time, the voltage drop between the installation earth electrode and the general mass of earth is determined by means of the voltage (potential) at terminals P1 and P2. The instrument then has sufficient information to calculate and display an ohmic value for the resistance of the installation earth electrode under test.

The temporary (current) electrode C2 is placed in the ground some distance away from the installation earth electrode under test. Unless the soil resistivity is very high, 20 metres is usually sufficient for this purpose. The temporary (potential) electrode P2 is placed approximately mid-way between the other two electrodes.

One reading is taken with P2 in the mid-way position, and a further two readings are taken with P2 approximately 2 metres on either side of the mid-way point.

The average of these three resistance readings should then be calculated, and compared with the individual readings. None of the individual readings should differ from the average by more than 5%. If they do, and there is a clear progression in the readings from one position of P2 to the next, it may mean that the resistance area of the installation earth electrode is overlapping that of temporary electrode C2. To overcome this problem, electrode C2 should be moved further away from the installation earth electrode, and the test procedure repeated. When three acceptable resistance readings have been obtained, record the average value on the certificate or report.

Re-connect the earthing conductor and all bonding and protective conductors on completion of the test, before the installation is energised (or re-energised).

Measurement of installation earth electrode resistance

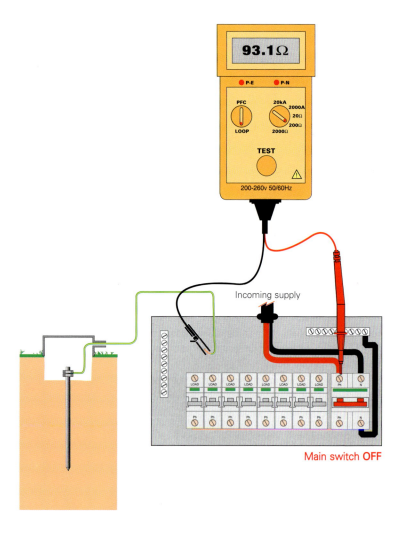

Method 2 (using an earth fault loop impedance test instrument).

Method 2

This method is essentially the same as the method used to measure the external earth fault loop impedance, Z_e. The measured value of Z_e for a TT system may therefore also be used as the approximate value of the resistance of the installation earth electrode.

The test method measures the impedance of the earth fault loop from the earthing conductor via the installation earth electrode, the earth return path, the source earth electrode, the supply distribution transformer, and the supply phase conductor back to the phase supply terminal at the origin of the installation.

As the impedance of the other parts of the earth fault loop can be expected to be relatively low, the measurement given by the earth fault loop impedance test instrument may be taken as the resistance of the installation earth electrode. (The measurement, which will include the resistance of the means of earthing at the source distribution transformer, will err on the safe (high) side).

Before carrying out the test by this method, the installation must be isolated, and the earthing conductor then disconnected from the main earthing terminal to remove any parallel paths to the general mass of earth provided by other protective conductors. As with Method 1, if such parallel paths are not removed before the measurement is made, a false (low) reading of the approximate resistance of the electrode to earth will be obtained.

The earth fault loop impedance test instrument is connected to the supply (energised) side of the main switch, and to the earthing conductor, as indicated in the figure opposite.

Re-connect the earthing conductor and all bonding and protective conductors immediately on completion of the test.

Why do we have to measure the resistance of the installation earth electrode?

BS 7430: Code of Practice for Earthing, suggests that a value of earth electrode resistance exceeding 100Ω may be unstable. In addition, Regulation 413-02-20 requires that the following condition is fulfilled for each circuit:

R_A multiplied by I_a is not to be more than 50 V $(R_A\ I_a \leq 50\ V)$

Where R_A is the sum of the resistance to earth of the electrode and the resistance of the protective conductor(s), including circuit protective conductor(s), connecting it to the exposed-conductive-part, and I_a is the current causing automatic operation of the protective device within 5 seconds.

Inspection, Testing, Certification and Reporting

When the protective device is a residual current device, I_a is the rated residual operating (tripping) current $I_{\Delta n}$ of the device. For certain special locations where there is an increased risk of electric shock, such as construction sites, agricultural and horticultural premises, Part 6 of *BS 7671* requires that the previous condition is replaced by:

$$R_A \, I_a \leq 25 \text{ V}$$

Knowledge of the resistance of the installation earth electrode is therefore essential in order to be able to verify that the installation satisfies these safety requirements.

Note: When this condition is satisfied, it does not indicate that the touch voltages under earth fault conditions will be limited to 50 V (or 25 V).

CONTINUITY OF PROTECTIVE CONDUCTORS

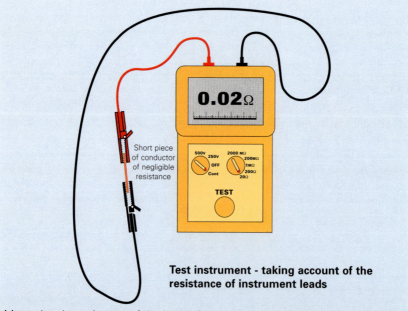

Test instrument - taking account of the resistance of instrument leads

Measuring the resistance of the leads of a continuity test instrument is necessary so that this value can be subtracted from test results for protective conductor continuity. Some test instruments have a built in facility to 'zero out' the lead resistance, making subtraction from test results unnecessary.

CONTINUITY OF PROTECTIVE CONDUCTORS

Errors caused by parallel paths

Before carrying out a test to confirm the continuity, or to measure the resistance, of a protective conductor, care should be taken to locate and disconnect (where practicable) any conductive paths which may be connected in parallel with all or part of it. Such parallel paths could be present due, for example, to a connection from a protective conductor to an electrically continuous system of pipework at one or more points along the protective conductor. Such parallel paths could lead to incorrect test results and may even conceal the fact that a protective conductor is not electrically continuous.

However, it is recognised that in some installations, such as those incorporating metal conduit or metal trunking which is fixed to a metallic structure, or where luminaires are fitted to structural steel, it would be impracticable to attempt to disconnect all parallel paths. In such circumstances, the continuity of a protective conductor may be verified only by inspection in addition to testing, during erection and after completion.

Test instrument

The test instrument to be used is an ohmmeter having a low ohms range, or an insulation and continuity test instrument set to the continuity range. It is recommended that the instrument has a short-circuit output current of not less than 200 mA (ac or dc) and an open-circuit voltage of between 4 V and 24 V. Continuity test readings of less than 1 Ω are common. Therefore, the resistance of the test leads is significant, and should not be included in any recorded test result. If the particular test instrument does not have provision for correcting for the resistance of the test leads, it will be necessary to measure the resistance of the leads when connected together, and for this measured value to be subtracted from all the test readings. See adjacent illustration.

The $R_1 + R_2$ method

This method is applicable to circuit protective conductors and their associated phase conductors. It is not applicable to main or supplementary bonding conductors. The $R_1 + R_2$ method of testing the continuity of circuit protective conductors has the following advantages:

- The information required to complete the $R_1 + R_2$ column of the Schedule of Test Results is obtained directly.

- Polarity is verified as each continuity test is performed.

- The method may be more convenient than the 'wander lead' method (see following section).

Testing continuity of the protective conductor to a lighting point by the $R_1 + R_2$ method

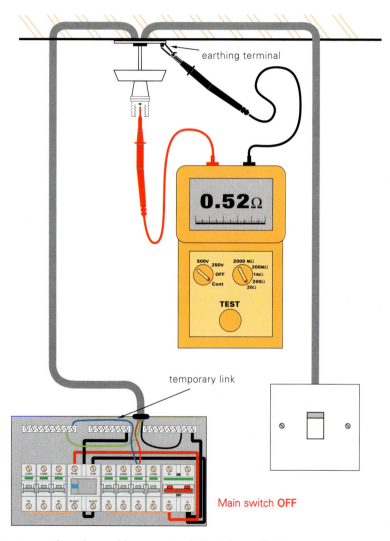

earthing terminal

0.52Ω

temporary link

Main switch OFF

This test procedure also enables correct polarity to be verified.

The test provides a measured value of resistance for the phase conductor and the circuit protective conductor of a circuit, in series. The test may be carried out, **having first securely isolated the supply**, by temporarily connecting together the phase conductor and circuit protective conductor at the supply end, and then connecting the test instrument to the terminals of the phase conductor and the circuit protective conductor at each point and accessory on the circuit in turn.

The value of $R_1 + R_2$ measured at the point or accessory electrically most remote from the supply will be the maximum value, and it is this value that should be recorded on the Schedule of Test Results. The test readings at the other points and accessories on the circuit serve to verify that the protective conductor is continuous to each of them, and that the polarity is correct.

Procedure:

- Securely isolate the supply.

- Disconnect any bonding connections that could affect the test readings.

- Make a temporary connection, at the supply end of the circuit (for example, at a consumer unit) between the phase and cpc of the circuit to be tested.

- Measure the resistance between phase and cpc at each point and accessory on the circuit (thereby also confirming correct polarity).

- Record the reading at the electrically most remote point or accessory (the highest value) of the circuit (and confirm correct polarity for the circuit).

- Re-connect any bonding conductors disconnected for the test.

- Remove the temporary connection.

Testing continuity of the protective conductor at a lighting switch by the wander lead method.

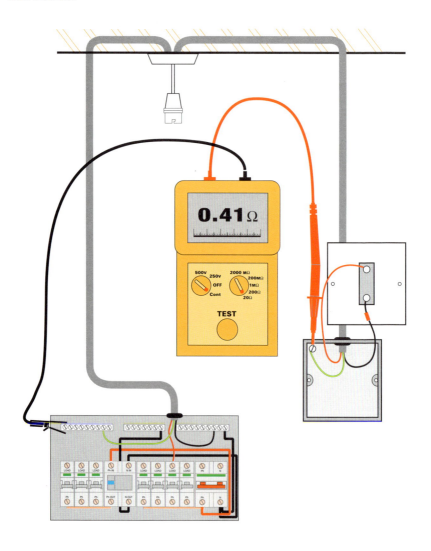

The wander lead method

This method is used principally for testing protective conductors that are connected to the main earthing terminal (main equipotential bonding conductors, circuit protective conductors and so on). The correct polarity of circuit connections will need to be verified separately.

Where a continuity test involves the opening of enclosures etc, that part of the installation will need to be isolated.

One terminal of the continuity test instrument is connected to the main earthing terminal with a long lead (or 'wander lead') and, with a lead from the other terminal, contact is made with the protective conductor at every position to which it is connected in that circuit, such as at socket-outlets, lighting points, fixed equipment points, switches, exposed-conductive-parts and extraneous-conductive-parts. By this means, provided that no parallel paths are present, the continuity of the protective conductor back to the main earthing terminal can be verified, and its resistance measured.

A main or supplementary bonding conductor can be tested by simply attaching the leads of the test instrument to each end of the conductor, having temporarily disconnected one end of that conductor to remove parallel paths.

Why do we need to test the continuity of protective conductors?

It is essential to ensure that all circuit protective conductors and bonding conductors are continuous. Otherwise, an exposed-conductive-part, an extraneous-conductive-part or the earthing terminal of a point or accessory could be left without an effective connection to earth, giving no protection against indirect contact in the event of an earth fault.

Whilst the primary purpose of testing for protective conductor continuity is to ensure that such continuity exists, the result of an $R_1 + R_2$ test can be used to determine the earth fault loop impedance at the point or accessory at which the test is applied. This procedure is explained under the subject heading of 'earth fault loop impedance' in this Chapter.

Use of high-current, low-impedance ohmmeter

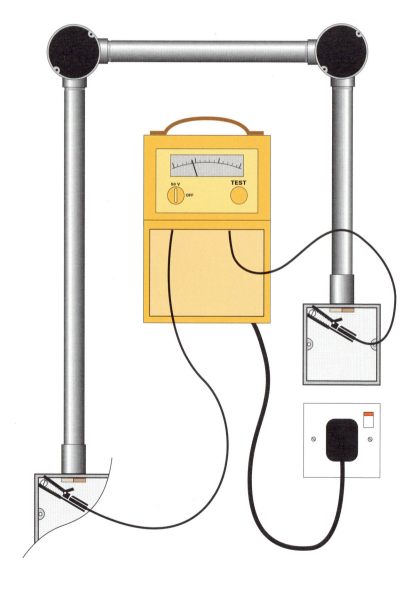

Continuity testing of ferrous enclosures

IEE Guidance Note 3 describes a special procedure for use where a ferrous enclosure such as steel conduit, steel trunking or steel wire cable armouring is used as a protective conductor. This procedure involves firstly performing a standard test by either the $R_1 + R_2$ or wander lead method, then inspecting the enclosure along its length to verify its integrity. If the inspector suspects the soundness of this conductor, the procedure continues with a further test using an earth fault loop impedance test instrument after connection of the supply. Should the inspector still suspect the soundness of the conductor at this stage, the procedure goes on to say that a further test may be performed using a high-current, low-impedance ohmmeter.

The high-current, low-impedance ohmmeter referred to in the procedure has a test voltage not exceeding 50 V and can provide a current approaching 1.5 times the design current of the circuit, although the procedure states that current need not exceed 25 A.

Where inspection and testing of ferrous enclosures used as protective conductors is undertaken, the inspector should make every effort to verify the compliance of these conductors with the requirements of *BS 7671* by careful inspection, so far as is reasonably practicable, and testing by the $R_1 + R_2$ or wander lead method. If, after this inspection and testing, the inspector has reasonable doubt about the soundness of any part of these conductive paths, such as the ability of any of the joints to provide durable electrical continuity and adequate mechanical strength, the inspector may proceed in one of two ways. The inspector may either decide to arrange for remedial action to be taken to correct or remove the defect (or recommend such action in the case of a periodic inspection report), or decide to follow the remainder of the procedure advocated in *IEE Guidance Note 3*, using an earth fault loop impedance test instrument and, if considered necessary, a high-current, low-impedance ohmmeter.

It is important to realise that the results of tests carried out using an earth fault loop impedance test instrument or a high-current, low-impedance ohmmeter cannot alone be regarded as proof of the soundness of a protective conductor. The inspector should consider this carefully before embarking on such tests. In addition, as pointed out in *IEE Guidance Note 3*, care needs to be taken when using a high-current tester, as sparking can occur at a faulty joint. This test should not be carried out if such sparking could cause danger (for example, by igniting combustible material, as might occur when undertaking periodic inspection in an unfamiliar property).

Ring final circuit continuity steps

Step 1

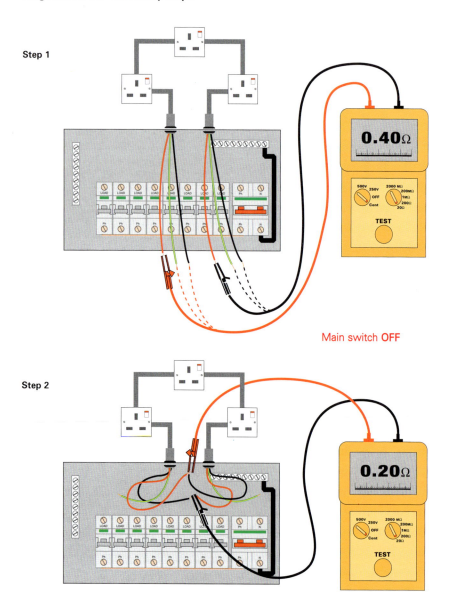

Main switch **OFF**

Step 2

CONTINUITY OF RING FINAL CIRCUIT CONDUCTORS

The cables of a ring final circuit start at the outgoing terminals of a consumer unit or distribution board, connect to all the points in the ring, and return to the same outgoing terminals. The live (phase and neutral) conductors must form a complete unbroken loop without interconnections, as must the circuit protective conductor, unless it is formed by a continuous metal covering or enclosure (such as a steel conduit) containing all the live conductors.

The test instrument to be used is one of the types described earlier in this Chapter for testing the continuity of protective conductors. The instrument needs to be capable of distinguishing between resistances differing by as little as $0.05\ \Omega$. As for the testing of protective conductors, due allowance needs to be made for the resistance of the test leads.

Procedure:

Step 1

- Securely isolate the installation, or the part of the installation to be tested.

- Identify and disconnect the phase, neutral and circuit protective conductors of the ring final circuit.

- Measure the resistance of the phase and neutral conductors separately, between ends. If the circuit protective conductor is a ring conductor not made up of metal conduit, trunking or cable covering, also measure its resistance.

- Note the readings obtained as:

    End to end resistance of the phase conductor = r_1
    End to end resistance of the circuit protective conductor = r_2
    End to end resistance of the neutral conductor = r_n.

The values of r_1, r_1 and r_n will indicate whether or not the conductors are continuous; the phase and neutral conductors should have equal resistances. If the circuit protective conductor has a smaller cross-sectional area than the phase and neutral, its resistance should be higher. In the case of a ring circuit having $2.5\,mm^2$ live conductors and a $1.5\ mm^2$ circuit protective conductor, the resistance of the cpc should be about 1.67 times that of the phase or neutral conductor.

If these conditions cannot be confirmed, then either the ring conductors have not been correctly identified, or there is a wiring defect in the ring circuit.

Ring final circuit continuity steps (continued)

Step 3

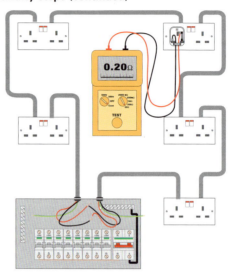

Step 4

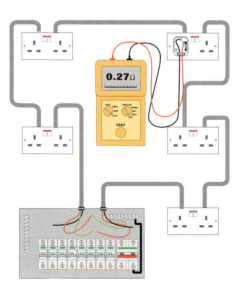

Step 2

- Connect the incoming neutral to the outgoing phase of the circuit, and vice versa, so that the conductors are 'cross-connected'.

- Measure the resistance (phase to neutral) between the pairs at the point at which the cross-connection is made, and note the result.

- The reading obtained should be approximately half that obtained for either the phase or neutral conductor in Step 1.

Step 3

With the circuit conductors still connected as in Step 2, measure between phase and neutral at each point on the ring (distribution board included). The reading should be substantially the same as in Step 2. Socket-outlets wired as spurs (cable branches) will give a slightly higher resistance reading depending on the length of the cable branch. Where the presence of a cable branch is identified from such a resistance reading, the inspector should assess whether adequate protection against overload has been afforded for that branch.

If the resistance measurements increase with the distance from the starting point, it is usually an indication that the ends of the ring have not been crossed over as intended. If this is the case, go back to the starting point and reverse the connections so that one phase is connected to the other neutral and vice versa, and then repeat the test.

It should now be found that the value is the same as in Step 2 at all points in the ring circuit.

Step 4

Repeat the procedure as in Steps 2 and 3, but using the phase and cpc conductors.

With the phase and cpc conductors cross-connected, the resistances measured between phase and cpc at each point on the ring will differ slightly from each other. Resistances measured at points connected to spurs will be higher, dependent on the length of the spur. The highest value obtained represents the maximum $R_1 + R_2$ value for the circuit, and can be recorded as such on the Schedule of Test Results. This value will be equal to $(r_1 + r_2)/4$, where r_1 and r_2 are the end-to-end resistances of the phase and circuit protective ring conductors respectively. This test sequence also verifies polarity and circuit protective conductor continuity to all points in the ring final circuit.

Step 5

Reconnect the conductors to the correct terminals.

Inspection, Testing, Certification and Reporting

Why do we need to check the continuity of ring final circuit conductors?

Incorrectly connected, or open circuit, conductors in an installed ring final circuit can lead to overloading of the circuit cables. Testing by the above procedure verifies continuity and correct connection, as well as correct polarity.

INSULATION RESISTANCE

Precautions may be necessary to avoid damage to electronic devices. The presence of any circuit or equipment vulnerable to a typical test should be indicated in the diagram or schedule provided for the installation. For practical reasons it may be necessary or desirable to sub-divide the installation in order to carry out insulation resistance testing.

Before proceeding with this test, the inspector should ensure that for the circuit to be tested:

- All pilot or indicator lamps that are likely to interfere with the test (by providing a path between phase and neutral) have been disconnected.

- All items of equipment which are at risk of damage from the test voltage have been disconnected.

- The incoming neutral has been disconnected so that there is no connection with earth.

The instrument to be used is an insulation resistance test instrument capable of supplying the test voltage indicated in Table 71A of *BS 7671* (which is reproduced below), when loaded with 1 mA.

Minimum values of insulation resistance:

Circuit nominal voltage (V)	Test voltage d.c. (V)	Minimum insulation resistance (MΩ)
SELV & PELV	250	0.25
Up to and including 500 V with the exception of the above systems	500	0.50
Above 500 V	1000	1.00

The insulation resistance is deemed satisfactory by *BS 7671* if the main switchboard and each distribution circuit tested separately with all its final circuits connected, but with current-using equipment disconnected, have an insulation resistance not less than the appropriate value given in Table 71A.

The insulation resistance is normally measured between live conductors (phase and neutral), and between live conductors and earth. Exceptionally, for periodic testing, if it is impracticable to disconnect electronic devices from a circuit, a measurement to protective earth only should be made, with the phase and neutral conductor connected together.

Insulation resistance testing between live conductors

To test the insulation resistance between live conductors (phase and neutral), a voltage is applied to the conductors, in pairs, from the origin of the circuit. Any leakage current from one conductor to the other is indicated in the form of a reduction in insulation resistance. Any short circuits, due to incorrect connections or damage to cables, are indicated by a low reading. Very low values of leakage current will be indicated by a resistance well beyond the range of the instrument, which may be 200 M Ω or more.

Procedure:

- Securely isolate the circuit under test from the supply, including the neutral.

- Disconnect all current-using equipment including fluorescent and other discharge luminaires. Remove filament lamps and check that all switches are closed (two-way switches should be operated to include all live conductors in the test). Exceptionally, where the disconnection of current-using equipment or removal of lamps is impracticable, the local switches controlling these items should be opened.

- Disconnect all equipment which is vulnerable to the test voltage.

- Check that the instrument and leads are in sound condition.

- Check the condition of the batteries in the test instrument.

- Select the appropriate test voltage and range.

- Connect the instrument, and record the readings as follows:

  - For a single-phase installation:

    Phase to neutral

  - For a three-phase installation:

    Phase 1 to Phase 2
    Phase 2 to Phase 3
    Phase 3 to Phase 1

    Then test each phase in turn to the neutral.

Measurement of the insulation resistance between live conductors and earth for a single-phase distribution board with all its final circuits connected, but with current-using equipment disconnected

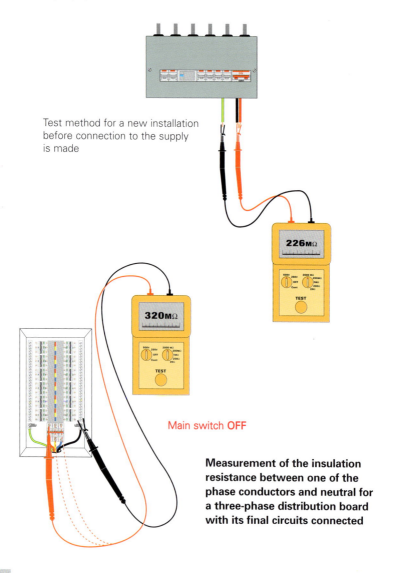

Test method for a new installation
before connection to the supply
is made

226MΩ

320MΩ

Main switch **OFF**

Measurement of the insulation resistance between one of the phase conductors and neutral for a three-phase distribution board with its final circuits connected

Insulation resistance between live conductors and earth

Procedure:

Having taken the necessary precautions as before, the inspector should:

- Link together all phase conductors and neutral.
- Connect the test instrument between the linked conductors and protective earth.
- Measure and record the result.

As may be seen from Table 71A of *BS 7671*, the minimum acceptable value for a 400/230 V installation may be as low as 0.5 MΩ. However, such a very low value should not be accepted without question. If any test reveals an insulation resistance lower than 2 MΩ, the circuits should be separated and tested individually in order to identify the cause of the low reading so that it can be remedied.

Why do we need to test insulation resistance?

The insulation resistance tests give an indication of the condition of the insulation of the installation. Effective insulation is required to provide protection against direct contact, and to prevent short-circuits and earth faults.

Unintended leakage currents due to inadequate insulation can present a risk of electric shock to persons and livestock. They can also cause further deterioration to the insulation and conductors if allowed to persist, when they may present a fire risk. It is therefore important that insulation resistances are tested either to verify the satisfactory condition of new installation work before the supply is connected, or to check the condition of an existing installation.

Inspection, Testing, Certification and Reporting

SITE-APPLIED INSULATION

Where insulation applied on site during the erection of the installation is relied upon for protection against direct contact, it has to be tested to verify that it is capable of withstanding, without breakdown or flashover, an applied voltage test equivalent to that specified in the British Standard for similar type-tested equipment.

Where site-applied supplementary insulation is relied upon for protection against indirect contact, it should be tested to ensure that the insulating enclosure provides a degree of protection not less than IP2X or IPXXB. The applied voltage test referred to above should also be carried out.

Because site-applied insulation is not commonly used, and applied voltage test procedures are specialised, no further information on the above matters is given in this book. Safety advice in relation to the high voltage used for testing should be obtained in the event of such testing being necessary.

PROTECTION BY SEPARATION OF CIRCUITS

Protection by separation of circuits is the generic term used for protection by SELV (separated extra-low voltage), PELV (protective extra-low voltage), and electrical separation. In each of these measures for protection against electric shock, the circuits and the source of supply have to provide electrical separation from the live parts of other systems, and (except where exempted by *BS 7671*) from the protective conductors of other systems.

To verify that the electrical separation meets the requirements of *BS 7671*, measurement of insulation resistance is generally necessary. Protection by separation of circuits is subject to detailed requirements given in *BS 7671*. Some of these requirements may be unfamiliar, especially those relating to electrical separation. Particular care is needed in the inspection and testing of installations which rely on these protective measures for safety.

Circuits

The instrument used for testing purposes must be capable of supplying the relevant dc test voltage when loaded with 1 mA. For SELV circuits a test of insulation resistance is carried out between the live conductors of each SELV circuit connected together and the protective conductors of any adjacent higher voltage system connected together. The test is applied at 250 V and the minimum acceptable insulation resistance is 0.25 MΩ. This test is unnecessary if the SELV circuits and other circuits are physically separated from each other throughout their length.

For PELV circuits, the tests are as for SELV, except that testing between the PELV circuits and earth is not required.

For electrical separation, a test of insulation resistance is carried out between the live conductors of each separated circuit connected together, and the protective conductors of each separated circuit (if any) and the live and protective conductors of any other circuits, connected together. The test voltage is 500 V and the minimum acceptable insulation resistance is 0.5 MΩ. This test is not necessary for a circuit which is physically separated from other circuits throughout its length.

Sources of supply

For a SELV or PELV system, the source of supply must be selected from one of those listed in Regulation 411-02-02 of *BS 7671*. This means that it must be derived from a safety isolating transformer complying with *BS EN 60742 (BS 3535)* in which there is no connection between the output winding and the transformer casing or the protective earthing circuit if any. Alternatively, the source may be one of the other listed types, giving equivalent electrical separation.

For electrical separation, the source of supply must be selected from those listed in Regulation 413-06-02, which means that an isolating transformer complying with *BS EN 60742 (BS 3535)* with no connection between its output winding and transformer casing or protective circuit, if any, must be used, or one of the other listed types giving equivalent electrical separation.

If the source of supply for a system is 'type-tested' to prove that electrical separation is provided to the standard required by the appropriate regulation of *BS 7671*, there is no need to carry out tests on the source for this purpose as part of the inspection and testing procedure for the installation. If, on the other hand, it cannot be established that the source has been suitably type-tested, it will be necessary to conduct tests to verify compliance with the above regulations. These tests, which are specialised and involve the use of high voltage, are beyond the scope of this book.

Relays and contactors

Any relays or contactors in electrically separated circuits must provide separation from any other circuits to the standard provided between the input and output of an isolating transformer to *BS EN 60742 (BS 3535)*. Unless such items are type-tested for electrical separation, it will be necessary for them to be tested in the manner described above for sources of supply.

Use of the standard test finger

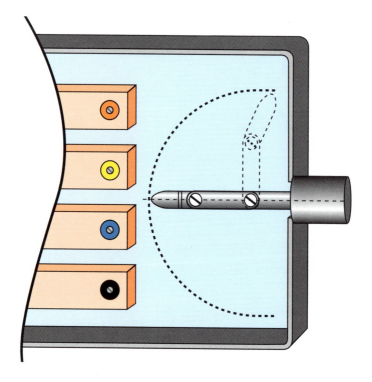

PROTECTION AGAINST DIRECT CONTACT BY BARRIER OR ENCLOSURE PROVIDED DURING ERECTION

This is a test to verify that each barrier or enclosure provided during erection gives suitable protection against the risk of electric shock through contact with live parts by fingers, solid objects and the like.

The test also applies to barriers or enclosures which, although provided in factory-built equipment, have been affected by the erection process. For example, if an opening has been formed in an enclosure on site for the entry of cables, but is oversized or unused, the enclosure must be capable of meeting the test criteria in order to comply with the requirements of *BS 7671*. If the enclosure fails to comply when tested, remedial action should be taken (such as reducing the size of the opening).

The degree of protection provided by each barrier or enclosure must not be less than IP2X or IPXXB. For readily accessible top surfaces the degree of protection must not be less than IP4X. IP2X means that the barrier or enclosure provides protection against contact with live parts by a standard test finger which is 12 mm in diameter, 80 mm long and is capable of bending through $90°$ twice in the same plane (like a normal finger). The test finger is applied with a force, not exceeding a specified maximum value and is used in conjunction with an electric signalling circuit. IPXXB is equivalent to IP2X in this context.

IP4X is protection against the entry of a wire, strip or similar object which is more than 1 mm thick, or a foreign object which is greater than 1 mm in diameter.

The test procedures are more fully described in *IEE Guidance Note No 3*. In practice, a visual inspection will usually be sufficient to confirm the compliance or otherwise of the barriers and enclosures with the above requirements. If all unused entries are suitably closed, a test may not be necessary.

INSULATION OF NON-CONDUCTING FLOORS AND WALLS

Testing the insulation of non-conducting floors and walls is required where 'protection by non-conducting location' is used.

This method of protection against indirect contact, although included in *BS 7671*, is not recognised in that Standard for general use. Protection by non-conducting location is only to be applied in special situations which are under effective supervision. The protective method is prohibited in certain installations and locations of increased shock risk as covered in Part 6 of *BS 7671*.

The testing procedures are specialised and beyond the scope of this book. The requirements are given in *BS 7671*, together with details of the stringent requirements for the installation, which should all be inspected to check compliance. Guidance is available in *IEE Guidance Note 3*.

Inspection, Testing, Certification and Reporting

POLARITY

Polarity tests are made to verify that any single-pole control devices or single-pole protective devices, are connected in the phase conductor only. This includes switches, fuses, thermostats etc. All non-reversible plugs and socket-outlets also need to be checked for correct connection. Edison screw and other centre-contact lampholders should be connected in such a way that the phase conductor is connected to the centre contact, and the neutral to the screw thread or the part of the lampholder which makes contact with the lamp cap. A check should also be made to ensure that the polarity of the incoming supply is correct, otherwise the whole installation would have the wrong polarity. (Such situations are not unknown).

The tests should be carried out before the circuit is energised, using a low-reading ohmmeter or continuity test instrument.

Much of the polarity testing can be carried out during the process of testing cpc continuity by using the $R_1 + R_2$ test method. If the results of such tests confirm correct polarity, there is no need to undertake polarity tests separately.

Why do we need to confirm correct polarity?

Incorrect polarity can give rise to danger in a number of ways, including:

- Parts of the installation may remain connected to the phase conductor when switched off by a single-pole device but, for all intents and purposes, will appear to be 'dead'.

- In the event of an overload, the circuit-breaker or fuse protecting that part of the installation would disconnect the neutral of the circuit, leaving the load at full phase voltage.

- An earth fault current might remain undetected by overcurrent protective devices.

Chapter 8

EARTH FAULT LOOP IMPEDANCE, Z_s

The measurement of earth fault loop impedance at the origin of the installation (Z_e) is described earlier in this Chapter. The guidance below relates to the measurement of earth fault loop impedances other than Z_e. The type of earth fault loop impedance test instrument to be used is the same as that used for the measurement of Z_e.

The earth fault loop impedance, Z_s, should be measured for every distribution circuit and final circuit having Earthed Equipotential Bonding and Automatic Disconnection of supply (EEBAD) as its method of protection against indirect contact. The measurement should be made at the point or accessory electrically furthest from the supply to the circuit. During periodic testing, if the continuity of the circuit protective conductor is not tested at each point and accessory, it may be necessary to measure Z_s at each such point and accessory.

Measured values of Z_s can be obtained wholly by means of an earth fault loop impedance test instrument. Alternatively, the $R_1 + R_2$ test results obtained during continuity testing may be used, in conjunction with test results obtained by use of the loop impedance test instrument for other parts of the installation. Both methods are discussed in this section.

The earth fault loop impedance test instrument uses the circuit voltage to pass a test current through the phase earth loop. This type of test can cause the unwanted operation of residual current devices or certain overcurrent devices, such as 6 A Type B circuit-breakers which may be in circuit. In such cases, it may not be possible to obtain a measured value of Z_s. Some instrument makers include features in their earth fault loop impedance test instruments aimed at avoiding some or all of these problems of unwanted tripping. Manufacturers should be consulted for details.

Determination of Z_s using $R_1 + R_2$ measured values

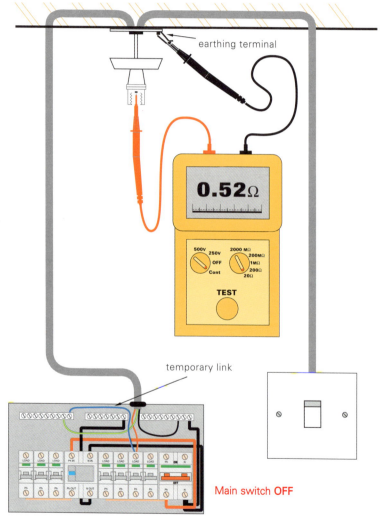

earthing terminal

0.52Ω

temporary link

Main switch **OFF**

$$Z_s = Z_e \text{ (or } Z_{db}) + (R_1 + R_2)$$

Determination of Z_s using $R_1 + R_2$ measured values

As previously indicated, an $R_1 + R_2$ test result for a given point in a circuit (generally the point which is electrically furthest from the supply) can be used to determine a measured value for the earth fault loop impedance, Z_s, at that point. To do this, it is also necessary to know the value of earth fault loop impedance, Z_{db}, at the distribution board or control panel etc directly supplying the circuit, generally from a measurement using an earth fault loop impedance test instrument (upstream of the circuit protective device if this is likely to be operated by the test). The following formula is then used:

$$Z_s = Z_{db} + (R_1 + R_2)$$

or, where the distribution board is at the origin of the installation:

$$Z_s = Z_e + (R_1 + R_2)$$

Strictly, it should not be $R_1 + R_2$ that is used in the above equations because this does not take account of the ac resistance or the inductive reactance of the phase conductor and circuit protective conductor. Instead it should be the impedance $Z_1 + Z_2$, which takes these quantities into account. Unfortunately, it is not possible to measure $Z_1 + Z_2$ using a dc test current, as supplied by most of the commonly available ohmmeters and continuity test instruments. To obtain a value for $Z_1 + Z_2$, it is necessary to use an ac test current at a frequency of 50 Hz in order to include the ac resistance and inductive reactance of the circuit conductors.

Fortunately, for circuits rated up to about 100 A, in which the phase and circuit protective conductors are part of the same wiring system or are immediately adjacent to each other and not separated by ferrous material (all as required by *BS 7671*), the ac resistance is not significantly greater than the dc resistance, and the inductive reactance is negligible. In such cases, $R_1 + R_2$ is approximately equal to $Z_1 + Z_2$, and the use of a test instrument supplying a dc test current will generally be sufficient.

For circuits rated at more than 100 A, the use of dc test current may lead to a value of Z_s which is optimistically low and may therefore be unacceptable. In such cases, an ohmmeter or continuity test instrument which supplies an ac test current at a frequency of 50 Hz should be used. Alternatively, a value of Z_s should be obtained by another method (such as direct measurement by an earth fault loop impedance test instrument).

Measurement of Z_s at a socket-outlet using an earth fault loop impedance test instrument

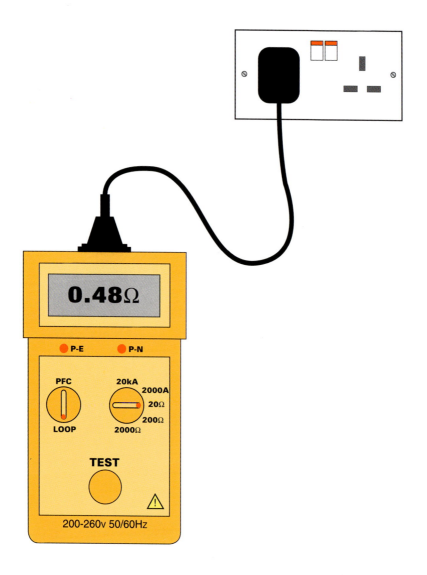

Procedure for measuring Z_s with a loop impedance test instrument

Testing should only be carried out after a thorough inspection and continuity testing have confirmed, so far as is reasonably practicable, that all circuit protective conductors, and main and supplementary bonding conductors are in place, and if the indicator lights on the instrument indicate that it is safe to proceed. If this procedure is not observed, there is a possibility that exposed-conductive-parts and extraneous-conductive-parts may become live during the testing process, thus exposing persons and livestock to the risk of electric shock.

- Remember that this test is conducted with the supply ON. All necessary precautions must therefore be taken to prevent danger.

- Care must be taken to minimise the risk of electric shock and/or burns when working on or near exposed live terminals. Connect the instrument securely before switching on the supply, perform the test, and then switch off the supply before disconnecting the instrument.

- Check that the test instrument, and all leads, clips and probes, are in good condition.

- Before pressing the button on the test instrument, check that the lamps or LEDs on the instrument indicate that it is safe to proceed. If they do not, investigate and resolve the situation before proceeding.

- For socket-outlets, as a minimum, perform the test on the outlet which is situated electrically furthest from the origin of the circuit. If this is not known, test all socket-outlets in the circuit and record the highest value. A proprietary test lead with a fitted plug should be used.

- For lighting circuits, as a minimum, Z_s at the electrically furthest point or accessory should be measured. This may be at a luminaire or a ceiling rose, but in some cases may even be at a switch.

The value of Z_s for every circuit must, after adjustment to take into account the heating effects of load current and possibly fault current, be sufficiently low to allow the overcurrent device to operate within the maximum time permitted by *BS 7671*, or should not exceed a value which might prevent conformity with the requirements of *BS 7671* relating to RCD-protected circuits.

Maximum earth fault loop impedance values for overcurrent protective devices in common use, for protection against indirect contact

MAXIMUM EARTH FAULT LOOP IMPEDANCE VALUES FOR OVERCURRENT PROTECTIVE DEVICES IN COMMON USE, FOR PROTECTION AGAINST INDIRECT CONTACT

For protection against indirect contact, the limiting values of earth fault loop impedances, Z_s, are given in Tables 41B1, 41B2 and 41D of BS 7671: 1992, for many commonly-used overcurrent protective devices.

The values given in those tables are the limits that apply under earth fault conditions, when the temperature of the conductors can be expected to be higher than when testing is undertaken (usually under no-load conditions). Consequently, the values of earth fault loop impedance when measured at ambient temperature should be lower than the limits set out in BS 7671.

It is generally accepted that, where the measured earth fault loop impedance of a circuit is not greater than 80% of the relevant limit specified in BS 7671, the impedance can be expected to be sufficiently low under earth fault conditions to meet the relevant limit specified in BS 7671, and for the protective device to automatically disconnect within the time specified.

The following table gives the limiting values of earth fault loop impedance when measured at ambient temperatures up to 20°C. The limits on measured values are 80% of the values given in BS 7671: 1992, rounded down. The boxes marked 'N/A' (Not Applicable) indicate either that the overcurrent protective device is not commonly available or that, by virtue of its characteristics, the device is not generally appropriate for protection against indirect contact.

The impedance values are based on the 'worst case' limits allowed by BS 7671 and, in certain cases, where the manufacturer of the protective device claims closer limits of fault current necessary for operation of the device than allowed for by the Standard, the values may be modified accordingly.

Where the measured value of the earth fault loop impedance exceeds the relevant tabulated value below, further investigation will be necessary to evaluate the particular circumstances to confirm that compliance with BS 7671 has been achieved.

LIMITING VALUES OF MEASURED EARTH FAULT LOOP IMPEDANCES
FOR COMMON OVERCURRENT PROTECTIVE DEVICES, FOR INDIRECT CONTACT, OPERATING AT 230 V
BASED ON 80% (APPROX.) OF THE VALUES GIVEN IN BS 7671: 1992

Nominal rating (A)	Fuses						Circuit-breakers to BS 3871 or BS EN 60898				
	BS 88 'gG' Parts 2 and 6		BS 1361		BS 3036		Type 1	Type 2	Type B	Type 3 and C	Type D
	0.4 s	5 s	0.4 s	5 s	0.4 s	5 s	0.4 s and 5 s				
5	N/A	N/A	8.72	13.68	8.00	14.80	9.60	5.48	N/A	3.84	1.92
6	7.11	11.28	N/A	N/A	N/A	N/A	8.00	4.56	6.40	3.20	1.60
10	4.26	6.19	N/A	N/A	N/A	N/A	4.80	2.74	3.84	1.92	0.96
15	N/A	N/A	2.74	4.17	2.13	4.46	3.20	1.83	N/A	1.28	0.64
16	2.25	3.48	N/A	N/A	N/A	N/A	3.00	1.71	2.40	1.20	0.60
20	1.48	2.43	1.42	2.34	1.48	3.20	2.40	1.36	1.92	0.96	0.48
25	1.20	1.92	N/A	N/A	N/A	N/A	1.92	1.09	1.53	0.76	0.38
30	N/A	N/A	0.96	1.53	0.91	2.20	1.60	0.91	N/A	0.64	0.32
32	0.87	1.53	N/A	N/A	N/A	N/A	1.50	0.85	1.20	0.60	0.30
40	0.68	1.12	N/A	N/A	N/A	N/A	1.20	0.68	0.96	0.48	0.24
45	N/A	N/A	0.48	0.80	0.49	1.32	1.06	0.60	0.85	0.42	0.21
50	0.50	0.87	N/A	N/A	N/A	N/A	0.96	0.55	0.76	0.38	0.19
60	N/A	N/A	N/A	0.58	N/A	0.93	N/A	N/A	N/A	N/A	N/A
63	0.38	0.68	N/A	N/A	N/A	N/A	0.76	0.43	0.60	0.30	0.15
80	N/A	0.48	N/A	0.41	N/A	N/A	0.60	0.34	0.48	0.24	0.12
100	N/A	0.35	N/A	0.30	N/A	0.44	0.48	0.27	0.38	0.19	0.09
125	N/A	0.28	N/A	N/A	N/A	N/A	N/A	N/A	N/A	N/A	N/A
160	N/A	0.21	N/A	N/A	N/A	N/A	N/A	N/A	N/A	N/A	N/A
200	N/A	0.16	N/A	N/A	N/A	N/A	N/A	N/A	N/A	N/A	N/A

This table is included with each pad of certificate and report forms published by the NICEIC.

Why do we need to test earth fault loop impedance?

Under earth fault conditions, a circuit which relies on EEBAD for protection against indirect contact is generally required to be disconnected under fault conditions within the maximum time permitted by *BS 7671* for that type of circuit or location.

Tables 41 B1, 41 B2 and 41D of *BS 7671* give maximum permitted values of Z_s for different types of overcurrent protective devices and different maximum permitted disconnection times. Shorter maximum disconnection times and modified regulations for RCD protection apply in certain special installations and locations including those covered by Part 6 of *BS 7671*.

As indicated above, it is generally necessary to adjust the values obtained by the test before comparing them with the maximum permitted values given in *BS 7671*. This is because the values referred to in *BS 7671* are based on the conductors having been heated up by the passage of load and/or fault current which increases their resistance, whilst test results are usually obtained when the conductor temperature is somewhat lower. Detailed advice on correction factors is given in IEE Guidance Notes 3 and 6. Alternatively, as a rule of thumb, the measured value of Z_s should not exceed 0.8 times the relevant value given in *BS 7671*, such as in Tables 41B and 41D, or in the equivalent tables in Part 6 of that Standard.

The table shown opposite gives the limiting values of measured earth fault loop impedance when measured at ambient temperatures up to 20°C. These limiting measured values are based on 80% of the values given in *BS 7671* for 5 second and 0.4 second disconnection, as appropriate. Where a shorter disconnection time is required, the limited values made be calculated as in the following example:

In an agricultural environment, a circuit is protected by a *BS 88*, 40 A fuse and is required to disconnect within 0.2 second. From Table 605B1 of *BS 7671*, the maximum permitted Z_s for the fuse is 0.71 Ω. By 'rule of thumb':

$$\text{Maximum permissible } Z_s \text{ test result} = 0.71 \text{ Ω} \times 0.80 = 0.56 \text{ Ω}$$

This means that the measured value of Z_s should not exceed 0.56 Ω for this circuit, in order to ensure disconnection within the maximum permitted time (in this case, 200 ms) in the event of an earth fault.

Where the measured value exceeds the 'rule of thumb' value of 0.56 Ω, further consideration must be given before accepting the measured value as being satisfactory.

Note that the value to be recorded in the schedule of test results is the measured value.

If the protective device is not of a type whose maximum permitted Z_s value is given in *BS 7671* then, as a rule of thumb, the measured value should not exceed 0.67 times the value given by the formula on the first page of Appendix 3 of *BS 7671*.

RCDs - conditions required for a satisfactory test result

General purpose RCDs to *BS 4293,* and RCD-protected socket-outlets to *BS 7288*

Test	Instrument test current setting	Satisfactory result
Test 1	100% of rated operating current	Device should trip in less than 200 ms

RCDs to *BS 4293* (incorporating a time delay)

Test	Instrument test current setting	Satisfactory result
Test 1	100% of rated operating current	Device should trip between 200 ms+ 50% of the time delay and 200 ms+ 100% of the time delay

For example, if the trip has a rated time delay of 100 ms, it should operate between 250 ms and 300 ms

General purpose RCDs to *BS EN 61008,* and RCBOs to *BS EN 61009*

Test	Instrument test current setting	Satisfactory result
Test 1	100% of rated operating current	Device should trip in less than 300 ms

RCDs to *BS EN 61008* and *BS 61009* Type 'S'

Test	Instrument test current setting	Satisfactory result
Test 1	100% of rated tripping current	Device should trip between 130 ms and 500 ms

OPERATION OF RESIDUAL CURRENT DEVICES

Each residual current device must be functionally tested to verify its effectiveness. The procedure comprises two basic parts, namely:

- Simulating an appropriate fault condition using an RCD test instrument.

- Testing the operation of the device by means of the integral test button.

Note: The test button on the RCD should not be operated until the electrical tests with the RCD test instrument have been completed, as this may influence the electrical test results.

Test instrument

The RCD test instrument should be capable of applying the full range of currents required to test the RCD in question, and of displaying the time, in milliseconds (ms), taken for the device to operate. For reasons of safety, the test instrument will normally automatically limit the duration of the test to a maximum of 2 seconds.

Testing of an RCD which incorporates an intentional time delay will only be possible if the test instrument is capable of providing a test duration which exceeds the permitted maximum operating time of the RCD. Owing to the variability of time delay, it is not possible to specify a maximum test duration, but 2 seconds should normally be found sufficient in practice.

Test procedure

- The RCD test instrument should not be used unless it has first been established that the earth fault loop impedance of the circuit is sufficiently low to satisfy the requirements of *BS 7671* for the RCD to be tested. That is to say, the loop impedance in ohms, when multiplied by the rated residual operating current of the device in amperes, must in general not exceed 50 V, or 25 V for certain special installations or locations as described in Part 6 of *BS 7671*.

- Contact by persons or livestock with exposed-conductive-parts or extraneous-conductive-parts of the installation must be prevented during the test. This is because potentially dangerous voltages may appear on these parts during the testing procedure.

- The test instrument is connected to the phase, neutral and earth on the load side of the RCD, with the load disconnected. The connection can be made at any convenient point in the circuit, such as a socket-outlet or, using a split 3-lead connection incorporating probes, at the outgoing terminals of the device. If the supply is 2 or 3-phase with no neutral connection available, the earth and neutral probes should both be connected to the protective earth. Check that the polarity indication on the instrument shows that it is in order to proceed with the test.

Testing of an RCD

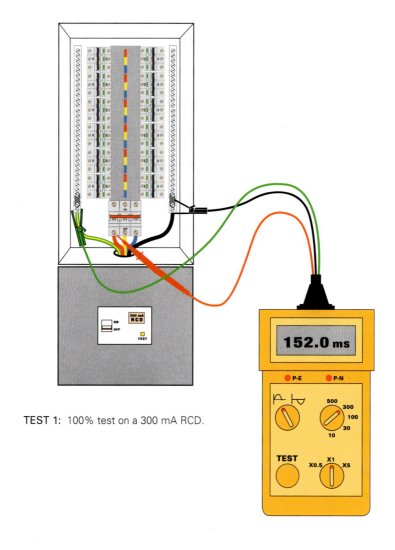

TEST 1: 100% test on a 300 mA RCD.

• Many RCD test instruments have a facility whereby tests can be carried out during the positive or negative half cycles of the supply waveform in turn. For tests 1 and 2 below, the operating time to be noted is the longer of those measured during the positive and negative half cycle tests.

Test 1

• Adjust the current setting to 100% of the rated operating (tripping) current of the device, push the button on the test instrument and note the time taken for the RCD to operate (See illustration opposite).

The measured operating time(s) should be compared with the appropriate value(s) in the tables shown on page 236, which relate to the type-tests carried out under specified conditions by RCD manufacturers in accordance with the relevant product standards.

Despite manufacturers' recommendations and the presence of the notice required by *BS 7671*, residual current devices may not have been operated by the users at quarterly intervals, by means of the 'test' button on the RCD. If, during a periodic inspection, a device does not trip within the expected time (including failing to trip at all) when first subjected to a test current equal to the rated operating (tripping) current, but operates satisfactorily when subjected to such a test subsequently (perhaps after manual switching), an appropriate observation and recommendation should be made on the report.

If the inspector considers that the result of the subsequent test is satisfactory and indicates the operating characteristic had the device been operated at quarterly intervals as intended, it would be in order to record the result of the subsequent test, subject to the related observation and recommendation. In particular, the attention of the recipient of the report should be drawn to the safety implications of not operating the device at quarterly intervals.

Testing of an RCD

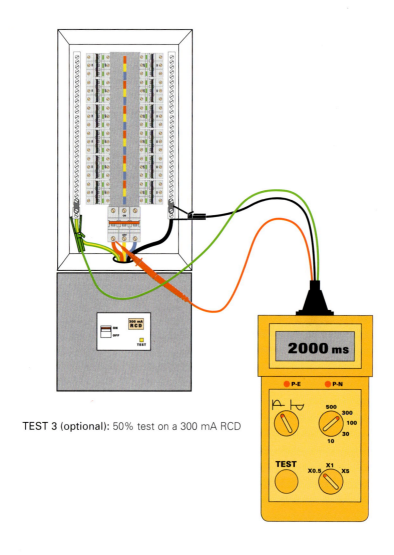

TEST 3 (optional): 50% test on a 300 mA RCD

Test 2

An RCD provided for supplementary protection against direct contact should have a rated residual operating (tripping) current not exceeding 30 mA. Regulation 412-06-02 requires that, when type-tested under specified conditions in accordance with the relevant product standards, such RCDs must operate in a time not exceeding 40 ms at a residual current of 150 mA.

When testing an installed RCD of this type using an RCD test instrument, the RCD should normally not be considered to be in a satisfactory condition unless it operates in a time not exceeding 40 ms when subjected to a residual (test) current of 150 mA.

It should be noted, however, that some RCDs intended to provide supplementary protection against direct contact are designed such that they operate under type-test conditions in a time only marginally less than 40 ms at a residual operating current of 150 mA. Such RCDs may operate under installed conditions in a time marginally exceeding 40 ms at 150 mA. If in doubt about the acceptability of measured RCD operating times, the inspector should consult the manufacturer to help determine whether the RCD(s) should be replaced.

Test 3 (optional)

- Adjust the current setting to 50% of the rated operating (tripping) current of the device. (See illustration opposite).
- Push the button on the test instrument and wait.
- The RCD should not operate within 2 seconds.

Note: Where there is no neutral connection available, and the neutral probe of the test instrument is connected to protective earth as previously described, there is a consequent increase in the residual current, which may cause the RCD to operate during **Test 3**. If this occurs, the manufacturer of the RCD should be consulted about the operating parameters of the device in order to establish whether it is or is not faulty.

The test button

Following the electrical test procedure described above, each RCD should be operated by means of its integral test facility. This confirms that the device is responding to its design level of sensitivity and that all the mechanical parts are functioning. The users of installations are advised (by means of a notice at or near the origin of the installations - see Regulation 514-12-02 of *BS 7671*) to carry out this simple test at quarterly intervals, but it is not a substitute for the electrical test procedure as described.

FUNCTIONAL TESTING OF ASSEMBLIES

All equipment should be checked to ensure that it functions as intended, and has been properly mounted, adjusted and installed in accordance with the relevant requirements of *BS 7671*.

Switching and controlling devices and systems should be checked for correct operation. Motors, motor drives and starters should be checked, along with lamps, luminaires and anything that forms part of the electrical installation. The inspector should not sign the Electrical Installation Certificate, the Domestic Electrical Installation Certificate, the Minor Electrical Installation Works Certificate or the Periodic Inspection Report until it has been established that all equipment has been properly mounted, adjusted and tested for satisfactory operation.

Where there is any doubt as to the intended method of controlling any part of an installation (such as a control system), the inspector should consult the designer of the installation (or, for an existing installation, the user, if appropriate).